Construction Research at NIOSH

Reviews of Research Programs of the National Institute for Occupational Safety and Health

Committee to Review the NIOSH Construction Research Program

Board on Infrastructure and the Constructed Environment

Division on Engineering and Physical Sciences

NATIONAL RESEARCH COUNCIL *AND*
INSTITUTE OF MEDICINE
OF THE NATIONAL ACADEMIES

THE NATIONAL ACADEMIES PRESS
Washington, D.C.
www.nap.edu

THE NATIONAL ACADEMIES PRESS • 500 Fifth Street, N.W. • Washington, DC 20001

NOTICE: The project that is the subject of this report was approved by the Governing Board of the National Research Council, whose members are drawn from the councils of the National Academy of Sciences, the National Academy of Engineering, and the Institute of Medicine. The members of the committee responsible for the report were chosen for their special competences and with regard for appropriate balance.

This study was requested by the National Institute for Occupational Safety and Health and supported by Contract Nos. 200-2000-00629 (Task Order #0033) and 200-2005-10881 (Task Order #0004) between the National Academy of Sciences and the National Institute for Occupational Safety and Health. Any opinions, findings, conclusions, or recommendations expressed in this publication are those of the author(s) and do not necessarily reflect the views of the organizations or agencies that provided support for the project.

International Standard Book Number-13: 978-0-309-12850-6
International Standard Book Number-10: 0-309-12850-1

Cover: Photos courtesy of Matt Gillen, NIOSH Construction Program.

Additional copies of this report are available from the National Academies Press, 500 Fifth Street, N.W., Lockbox 285, Washington, DC 20055; (800) 624-6242 or (202) 334-3313 (in the Washington metropolitan area); Internet, http://www.nap.edu.

Printed in the United States of America

Suggested citation: National Research Council and Institute of Medicine. 2009. *Construction Research at NIOSH.* Committee to Review the NIOSH Construction Research Program. Rpt. No. 8, Reviews of Research Programs of the National Institute for Occupational Safety and Health. Washington, D.C.: The National Academies Press.

THE NATIONAL ACADEMIES

Advisers to the Nation on Science, Engineering, and Medicine

The **National Academy of Sciences** is a private, nonprofit, self-perpetuating society of distinguished scholars engaged in scientific and engineering research, dedicated to the furtherance of science and technology and to their use for the general welfare. Upon the authority of the charter granted to it by the Congress in 1863, the Academy has a mandate that requires it to advise the federal government on scientific and technical matters. Dr. Ralph J. Cicerone is president of the National Academy of Sciences.

The **National Academy of Engineering** was established in 1964, under the charter of the National Academy of Sciences, as a parallel organization of outstanding engineers. It is autonomous in its administration and in the selection of its members, sharing with the National Academy of Sciences the responsibility for advising the federal government. The National Academy of Engineering also sponsors engineering programs aimed at meeting national needs, encourages education and research, and recognizes the superior achievements of engineers. Dr. Charles M. Vest is president of the National Academy of Engineering.

The **Institute of Medicine** was established in 1970 by the National Academy of Sciences to secure the services of eminent members of appropriate professions in the examination of policy matters pertaining to the health of the public. The Institute acts under the responsibility given to the National Academy of Sciences by its congressional charter to be an adviser to the federal government and, upon its own initiative, to identify issues of medical care, research, and education. Dr. Harvey V. Fineberg is president of the Institute of Medicine.

The **National Research Council** was organized by the National Academy of Sciences in 1916 to associate the broad community of science and technology with the Academy's purposes of furthering knowledge and advising the federal government. Functioning in accordance with general policies determined by the Academy, the Council has become the principal operating agency of both the National Academy of Sciences and the National Academy of Engineering in providing services to the government, the public, and the scientific and engineering communities. The Council is administered jointly by both Academies and the Institute of Medicine. Dr. Ralph J. Cicerone and Dr. Charles M. Vest are chair and vice chair, respectively, of the National Research Council.

www.national-academies.org

COMMITTEE TO REVIEW THE NIOSH CONSTRUCTION RESEARCH PROGRAM

RICHARD L. TUCKER, *Chair,* University of Texas, Austin
PAUL BARSHOP, Independent Project Analysis, Ashburn, Virginia
MARIA BRUNETTE, University of Massachusetts, Lowell
PATRICIA A. BUFFLER, University of California, Berkeley
ANGELA DiDOMENICO, Liberty Mutual Research Institute for Safety, Hopkinton, Massachusetts
BRADLEY EVANOFF, Washington University, St. Louis, Missouri
LINDA M. GOLDENHAR, University of Cincinnati, Ohio
WILLIAM H. KOJOLA, AFL-CIO, Washington, D.C.
EMMITT J. NELSON, Nelson Consulting, Inc., Houston, Texas
PETER PHILIPS, University of Utah, Salt Lake City
IRIS D. TOMMELEIN, University of California, Berkeley

Staff

LYNDA STANLEY, Study Director
KEVIN LEWIS, Program Officer
DANA CAINES, Financial Associate
HEATHER LOZOWSKI, Financial Associate

Preface

Construction is unique among U.S. industries, annually producing buildings and infrastructure valued at more than $1.2 trillion. Industry practitioners characterize the construction industry as having four distinctly different sectors—residential, commercial, industrial, and heavy construction—with specialty trade contractors (e.g., carpenters, plumbers) involved in all four sectors. The sector differences are significant because they affect the implementation of worker safety and health programs.

The residential sector is the largest of the four sectors, but also the least organized, with millions of small contractors and a relatively unstructured craft environment. The commercial buildings sector is characterized by specialized subcontractors, with more highly trained workers grouped according to recognized building trades. The industrial sector is about the same size as the commercial buildings sector; it involves generally larger construction firms, operating in a direct-hire mode, with highly skilled workers and coordinated safety and health efforts. The heavy-construction sector, which builds roads, bridges, and other infrastructure, is more equipment oriented, less labor intensive, and primarily involves public-sector owners.

Buildings, structures, and infrastructure—the products of construction projects—last 20 to 100 years or more. In contrast, construction projects and the industry itself can be described as "temporary." Projects are built within several years; they may be located anywhere in the country. Organizations and personnel involved with a construction project change continuously: Individual organizations and personnel may not have worked together previously, and they may come from many different backgrounds and cultures. In the residential and commercial sectors particularly, crafts- and tradespeople are likely to work at more than one site or project during any given week or other time frame.

The unique nature of construction activities has resulted in the formation of specialized trades. The members of 1 of about 15 normally recognized building trades—for example, masons—have highly specialized but varied skills that have been developed over years of training and apprenticeship. Thus, construction craftspersons are in many ways highly knowledgeable artisans who are respected not only for their manual skills but also for their technical knowledge related to their specific crafts and to other interfacing crafts.

The uniqueness of the construction industry presents a challenge for occupational safety and health protection. The work environment is inherently less safe during construction than it is after construction is completed: For example, stairways and handrails are much safer after completion than during installation. Moreover, the work environment changes daily for individual workers as construction progresses, and the workers themselves change as different crafts are called in while a project is being built. Worker exposure to hazardous environments is difficult to track because workers move from project to project or company to company during their careers.

For all of these reasons, conducting research intended to improve the health and safety of construction workers is challenging. Empirical data for work-related illnesses and diseases are particularly difficult to gather owing to the temporary nature of the work and the latent nature of health and musculoskeletal disorders. Equally difficult is finding ways to transfer research results into workplace behaviors and practices that reduce fatalities, injuries, and illnesses at the worksite.

Nonetheless, it appears that significant progress has been made in reducing construction-related fatalities and injuries in recent years. The Occupational Safety and Health Administration and the Bureau of Labor Statistics have provided standard definitions for "Total Recordable Incident Rates" and "Days Away from Work Injuries" that, along with reporting requirements, have allowed meaningful tracking of job site injury rates. These rates have declined, possibly by a factor of two, over the past 15 years. The fatality rate attributed to construction incidents has also declined significantly. Much of this improvement can arguably be attributed to the leading industry companies and worker organizations, along with the support of construction equipment manufacturers, which have made concerted and organized efforts to provide safer workplaces. Increasing health care costs may have helped to spur greater attention to the prevention of injuries. In some industry sectors the culture has changed from contending that "construction is inherently dangerous—accidents happen" to holding that "zero accidents are achievable."

These improvements can also be attributed in part to the research and activities of the National Institute for Occupational Safety and Health's (NIOSH's) Construction Research Program.

In 2004, NIOSH asked the National Academies to review up to 15 specific NIOSH research programs to determine their relevance to and impact on various industries in the United States. The Committee to Review the NIOSH Construction Research Program was composed of persons with widely diverse backgrounds who have worked in academia, government, industry, and labor unions (Appendix D). To complete its tasks, the committee worked diligently, convening a series of meetings (Appendix B) and also employing conference calls and e-mail correspondence. The NIOSH staff provided useful and complete information and was available to answer committee questions. The National Research Council (NRC) staff provided invaluable assistance in gathering information and arranging interviews with industry representatives, in accordance with committee requests.

This report is reflective of many months of intense effort by the committee, the NIOSH staff, and the NRC staff. The committee's ratings for the relevance and impact of the Construction Research Program are made in the context of the program's limited resources, the segmentation of the industry, and other factors beyond the control of the program itself.

Chapter 1 of this report describes the background of the study and provides context for the committee's evaluation. Chapter 2 describes the NIOSH Construction Research Program and external factors that influence the capacity of the program to meet its goals and objectives. Chapter 3 describes activities undertaken by the Construction Research Program related to the four major research goals that the program focused on during the period reviewed for this study (between 1996 and 2005); it provides the committee's detailed assessment, evaluation, and ratings with respect to the program's relevance and impact in reducing workplace fatalities, injuries, and illnesses. Chapter 4 contains the committee's recommendations regarding areas of future research and program improvement.

Despite many obstacles, the committee believes that the NIOSH Construction Research Program has been highly relevant and has made important contributions to the reduction of fatalities, injuries, and illnesses at construction worksites. The committee hopes that this report will provide valuable guidance to NIOSH as it structures its Construction Research Program for the next decade.

Richard L. Tucker, *Chair*
Committee to Review the NIOSH
Construction Research Program

Acknowledgments

This report has been reviewed in draft form by individuals chosen for their diverse perspectives and technical expertise, in accordance with procedures approved by the National Research Council's Report Review Committee. The purpose of this independent review is to provide candid and critical comments that will assist the institution in making its published report as sound as possible and to ensure that the report meets institutional standards for objectivity, evidence, and responsiveness to the study charge. The review comments and draft manuscript remain confidential to protect the integrity of the deliberative process. We wish to thank the following individuals for their review of this report:

John C. Bailar III, University of Chicago,
Amit Bhattacharya, University of Cincinnati,
Bryan Buchholz, University of Massachusetts, Lowell,
John Dement, Duke University,
Bryan Drennan, Sandia National Laboratories,
Richard Dwyer, United Brotherhood of Carpenters,
Linda Forst, University of Illinois, Chicago,
Jimmie Hinze, University of Florida,
Gerry Lipka, Sandia National Laboratories,
James Melius, New York State Laborers Health and Safety Fund,
John Rosecrance, Colorado State University,
David H. Wegman, University of Massachusetts, Lowell, and
Wm. A. Wulf, University of Virginia.

Although the reviewers listed above have provided many constructive comments and suggestions, they were not asked to endorse the conclusions or recommendations, nor did they see the final draft of the report before its release. The review of this report was overseen by Richard Wright, former director of the Building and Fire Research Laboratory of the National Institute of Standards and Technology. Appointed by the National Research Council, he was responsible for making certain that an independent examination of this report was carried out in accordance with institutional procedures and that all review comments were carefully considered. Responsibility for the final content of this report rests entirely with the authoring committee and the institution.

The committee also acknowledges and appreciates the contribution of the members of the Board on Infrastructure and the Constructed Environment of the National Research Council (Appendix C). The board members were not asked to endorse the committee's conclusions or recommendations or to review the final draft of the report before its release.

Contents

APPENDIXES

Tables, Figures, and Boxes

TABLES

FIGURES

BOXES

Abbreviations and Acronyms

ABLES	Adult Blood Lead Epidemiology and Surveillance
AFL-CIO	American Federation of Labor and Congress of Industrial Organizations
ANSI	American National Standards Institute
BAT	Bureau of Apprenticeship Training
BLS	Bureau of Labor Statistics
BRDPI	Biomedical Research and Development Price Index
CD	compact disc
CDC	Centers for Disease Control and Prevention
CEI	Construction Ergonomics Initiative
COSH	National Council for Occupational Safety and Health
CPWR	Center to Protect Workers' Rights
CSC	Construction Steering Committee (NIOSH)
DARTs	days away from work, days of restricted work activity or job transfer
EC	Evaluation Committee
eLCOSH	electronic Library of Construction Occupational Safety and Health
EPA	Environmental Protection Agency

FACE	Fatality Assessment and Control Evaluation
FTE	full-time equivalent
FY	fiscal year
GDP	gross domestic product
HUD	Department of Housing and Urban Development
ITCP	Internal Traffic Control Plan
IUOE	International Union of Operating Engineers
MSD	musculoskeletal disorder
NAICS	North American Industry Classification System
NCC	National Construction Center
NIH	National Institutes of Health
NIOSH	National Institute for Occupational Safety and Health
NORA	National Occupational Research Agenda
NORA1	first National Occupational Research Agenda (1996-2005)
NORA2	second National Occupational Research Agenda (2005 forward)
NRC	National Research Council
OHPL	overhead power line
OSHA	Occupational Safety and Health Administration
PATH	Posture, Activity, Tools, and Handling
PWS	proximity-warning system
R2P	research-to-practice
RFA	request for applications
SSVR	surround-screen virtual reality
TRIR	Total Recordable Incident Rate

Summary

ABSTRACT *The construction industry is one of the largest sectors of the U.S. economy, accounting for about 13 percent of the gross domestic product and employing about 11 million workers in 2005. Construction is also one of the most dangerous industries for workers. Hazards for construction workers include working at heights, in excavations and tunnels, on highways, and in confined spaces; exposure to high levels of noise, to chemicals, and to high-voltage electric lines; and the use of power tools and heavy equipment. Significant health risks include hearing loss, silicosis,*[1] *musculoskeletal disorders, skin diseases, and health effects associated with exposures to lead, asphalt fumes, and welding fumes.*

The National Institute for Occupational Safety and Health (NIOSH) has conducted construction-relevant research activities since the 1970s. In 1990, Congress directed NIOSH to develop a comprehensive prevention program directed at health and safety problems affecting construction workers by expanding existing NIOSH activities in areas of surveillance, research, and intervention. During the period studied, 1996 through 2005, the NIOSH Construction Research Program focused on four research goals: reducing the major risks associated with traumatic injuries and fatalities, reducing exposure to

[1]Silicosis is a disabling and sometimes fatal lung disease caused by breathing dust that has very small pieces of crystalline silica in it. Crystalline silica is found in concrete, masonry, sandstone, rock, paint, and other abrasives.

health hazards, reducing major risks associated with musculoskeletal disorders, and increasing the understanding of construction industry attributes and factors for improving health and safety outcomes.

In conjunction with planned reviews of up to 15 NIOSH research programs, the Division on Engineering and Physical Sciences of the National Research Council convened a committee of experts to evaluate the relevance and impact of the NIOSH Construction Research Program. The committee evaluated the relevance of the program in terms of its research priorities and its connection to improvements in the protection of workers in the workplace; it evaluated the impact of the program in terms of its contributions to worker safety and health. The committee was also asked to assess the program's identification and targeting of new research areas, to identify emerging research issues, and to provide advice on ways that the program might be strengthened.

NIOSH cannot, on its own, set and enforce research-based standards or practices for the construction industry. These efforts are carried out, respectively, by the Occupational Safety and Health Administration (OSHA) and by individual contractors, unions, and other entities. Nonetheless the Construction Research Program can be expected to contribute to reductions in construction workplace fatalities, injuries, and illnesses through its research, dissemination of its research results, and transfer to practice. Taking into account a range of external factors beyond the control of the NIOSH Construction Research Program, the committee found that the program has indeed made meaningful contributions to improving construction worker safety and health.

The review by this committee was guided by a methodology and framework established previously by a National Academies parent committee (see Appendix A). Using a 5-point scoring scale (with 5 being the highest), the present committee assigned the Construction Research Program a score of 5 for relevance, indicating that the research is in high-priority areas and that NIOSH is significantly engaged in appropriate transfer activities for completed research projects and reported program results. Regarding impact, the committee assigned the program a score of 4, indicating that the research program has made some contributions to end outcomes (worker safety and health) or well-accepted intermediate outcomes.

To ensure the continued high level of relevance for the program's research and to enhance the impact of that research on health and safety practices within the construction industry, the committee recommends that research-to-practice (R2P) efforts involve individuals trained in or having the experience and skills to create strategic diffusion and social marketing plans for NIOSH research and to evaluate such plans' effectiveness; that consideration be given to having the majority of R2P efforts conducted through the National Construction Center; that high-level attention by NIOSH leadership be given to determining how to provide program resources that are commensurate with a more robust pursuit of the program's goals; that the positions of Construction Program Coordinator and Construction Program Manager both become full-time positions; that the National Construction Center continue to be used as an important component in NIOSH's Construction Research Program; and that the program establish a closer connection with OSHA and other regulatory and consensus standards organizations that can ensure that the program's research is applied effectively in rule-making efforts.

BACKGROUND AND STATEMENT OF TASK

Between 1992 and 2005, 16,000 construction workers in the United States died from work-related injuries. In 2005 alone, 1,243 construction workers died from job-related traumatic injuries. This number accounted for 22 percent of job-related deaths across all industries, a figure that is disproportionately high given that construction workers account for about 8 percent of the total workforce. The death rate for construction workers was almost three times that of full-time workers for all industries: 11.1 deaths per 100,000 construction workers compared with 4.2 deaths per 100,000 workers in all industries (CPWR, 2007, Section 32). Among all sectors, construction had the fourth highest rate of fatalities in 2005 (after agriculture, mining, and transportation) and the second highest rate of nonfatal injuries and illnesses (after transportation) as measured by days away from work: 239.5 per 10,000 construction workers compared with 135.7 per 10,000 workers for all industries (CPWR, 2007, Section 32).

Nonetheless, the trends in construction workplace safety show significant improvements. Between 1992 and 2005, the overall rate of construction-related fatalities declined from 14.3 to 11.1 per 100,000 workers, which translates to 350 fewer deaths per year for a workforce of 11 million. The rate of nonfatal injuries and illnesses also declined significantly, although the absolute rate of decline is not clear owing to changes in reporting requirements.[2] An additional complicating

[2]In 2002, the Occupational Safety and Health Administration implemented a number of changes in the definitions of injury and illness cases recorded by employers. The new definitions in turn resulted in changes in occupational injury and illness statistics provided by the Bureau of Labor Statistics. There is some disagreement as to the overall effect of these changes. For example, one author writes that "while these data follow the trend of declining cases and rates seen throughout the past decade, because of the change in definition they cannot be compared with data from prior years" (Wiatrowski, 2004). In contrast, others note that although the changes in coding systems have signifi-

factor is that a large portion of the construction sector is not reflected in statistics on nonfatal injuries and illness, because many workers are self-employed or work in construction firms that employ fewer than 10 workers, categories for which such statistics are not collected.

The National Institute for Occupational Safety and Health (NIOSH) was established by the Occupational Safety and Health Act of 1970 (Public Law 91-596) to "conduct (directly or by grants or contracts) research, experiments, and demonstrations relating to occupational safety and health, including studies of psychological factors involved, and relating to innovative methods, techniques, and approaches for dealing with occupational safety and health problems." The law also created the Occupational Safety and Health Administration (OSHA) to set and enforce standards for workplace safety and health and to work with employers and employees through technical assistance and consultation programs.

NIOSH, unlike OSHA, is not authorized to set and enforce standards. Instead, NIOSH as a research organization is authorized to carry out the following:

- Develop recommendations for occupational safety and health standards,
- Conduct research on worker safety and health,
- Conduct training and employee education,
- Develop information on safe levels of exposure to toxic materials and harmful physical agents and substances,
- Conduct research on new safety and health problems,
- Conduct on-site investigations (health hazard evaluations) to determine the toxicity of materials used in workplaces, and
- Fund research by other agencies or private organizations through grants, contracts, and other arrangements (CFR, 2008).

NIOSH has conducted research on health and safety hazards in the construction industry since the 1970s. The NIOSH Construction Research Program was formally established in 1990 after Congress directed NIOSH to develop a comprehensive prevention program focused on health and safety problems affecting construction workers by expanding existing NIOSH activities in the areas of surveillance, research, and intervention.

In September 2004, NIOSH requested that the National Academies conduct evaluation reviews of up to 15 specific NIOSH research programs to assess the relevance and impact of the work of NIOSH in reducing workplace injury and ill-

cantly affected the compatibility of injury and illness data for construction subsectors over time, the impact on the construction industry as a whole is relatively small (CPWR, 2007, Section 32).

ness. For consistency across the set of evaluations, each review uses a methodology and framework developed by the National Academies' Committee for the Review of NIOSH Research Programs (called the Framework Committee). The conduct of these reviews has been guided by the Framework Document presented in Appendix A.

In 2007, the National Research Council appointed the Committee to Review the NIOSH Construction Research Program, composed of experts from a wide range of disciplines who have worked in industry, academia, government, and labor unions (Appendix D provides biosketches of the committee members). The committee was tasked with reviewing NIOSH's Construction Research Program and evaluating the program's relevance and impact. The committee evaluated the relevance of the program in terms of its research priorities and its connection to improvement in the protection of workers in the workplace; it evaluated the impact of the program in terms of its contributions to worker safety and health. The committee was also asked to assess the program's identification and targeting of new research areas, to identify emerging research issues, and to provide advice on ways that the program might be improved. The committee chose to focus its review on the time period from 1996 through 2005.

NIOSH CONSTRUCTION RESEARCH PROGRAM

The mission of the NIOSH Construction Research Program is to eliminate occupational diseases, injuries, and fatalities among individuals working in the construction industry through a focused program of research and prevention (NIOSH, 2007).

Between 1996 and 2005, the program focused on four general research goals:

- Goal 1: Reduce the major risks associated with traumatic injuries and fatalities in construction.
- Goal 2: Reduce exposures to health hazards associated with major risks of occupational illness in construction.
- Goal 3: Reduce the major risks associated with musculoskeletal disorders in construction.
- Goal 4: Increase understanding of construction-sector attributes that affect occupational safety and health outcomes.

Two to six sub-goals are associated with each of the four major goals. The goals are a composite of goals and priorities that draw from the first National Occupational Research Agenda (NORA1) and internally generated strategic goals and high-priority topics.

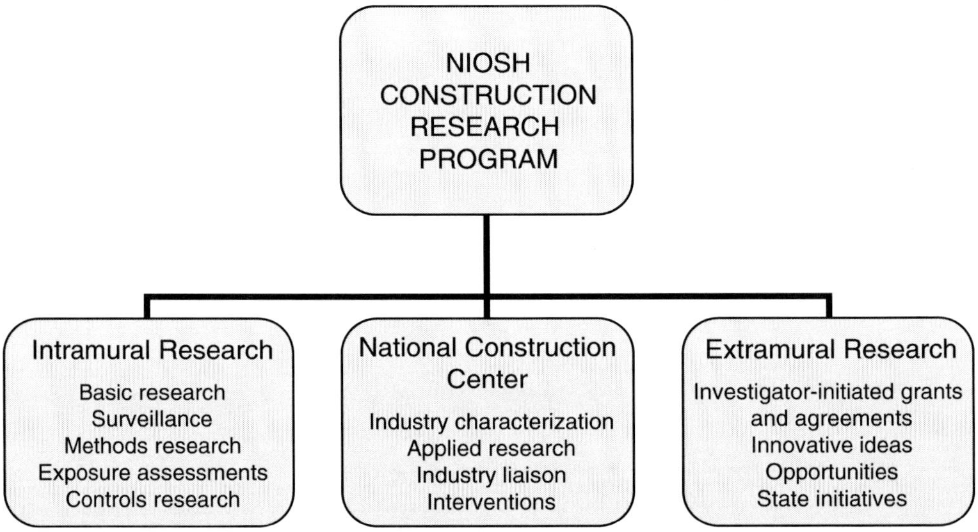

FIGURE S.1 Components of the NIOSH Construction Research Program.

To achieve its goals, the program uses a three-pronged structure to conduct research and disseminate research results: NIOSH-wide intramural research and surveillance programs, a National Construction Center cooperative agreement, and support grants and agreements for investigator-initiated extramural research projects (Figure S.1):

- *Intramural research* is carried out by in-house researchers assigned to six divisions and associated laboratories throughout NIOSH, rather than by a distinct construction research organizational entity. As such, the program operates as a matrix organization within NIOSH. Activities that focus on basic research, surveillance, methods research, exposure assessments, and controls research are managed through a coordinator and the Construction Steering Committee (CSC). Composed of representatives from each NIOSH division and laboratory,[3] the CSC briefs NIOSH senior executives on issues relating to the Construction Research Program.

[3]Applied Research and Technology; Surveillance, Hazard Evaluations, and Field Studies; Education and Information; Respiratory Disease Studies; Safety Research; Health Effects Laboratory; Pittsburgh Research Laboratory; Spokane Research Laboratory; Office of Extramural Programs; and National Personal Protective Technology. NIOSH's laboratories are located in Pittsburgh, Pennsylvania; Spokane, Washington; Cincinnati, Ohio; and Morgantown, West Virginia.

- *The National Construction Center* (NCC) is operated under a competitively awarded (scientifically and programmatically reviewed) 5-year cooperative agreement.[4] The NCC focuses on industry characterization, applied research, creating liaisons with the industry, and developing research-based interventions. The research is conducted by NCC staff dedicated to construction research, and through a consortium of universities whose staff conduct research through contracts with the NCC.
- *Extramural research* is conducted through investigator-initiated extramural grants and cooperative agreements (all scientifically and programmatically reviewed), and through support for state health department investigators working on construction health and safety surveillance and state-level initiatives.

Research conducted in any one component of the program is leveraged by way of interactions with researchers in the other two components during regularly scheduled meetings, construction conferences, and other construction-specific networking opportunities.

Total annual funding for the program between fiscal year (FY) 1997 and FY 2007 (which overlaps the period studied) averaged about $17.8 million, ranging from a high of $20.3 million in FY 1997 to a low of $13.8 million in FY 2007. When adjusted for inflation and changes in technologies, the funding level for the program has declined in terms of real purchasing power (NIOSH, 2007). Program funding levels have also declined as a portion of the total budget for NIOSH.

STUDY PROCESS

The National Academies' reviews of all NIOSH research programs are being conducted using a methodology and framework described in the Framework Document (Appendix A). Inputs to the review of the Construction Research Program included information provided by NIOSH program staff in oral presentations, a 500-page evidence package (NIOSH, 2007), and written responses to committee questions. The committee also received input from program stakeholders, including representatives from labor, industry, regulatory agencies, professional organizations, and academia (Appendix B). Individual committee members conducted research independently and also shared their collective expertise.

---

[4]The NCC agreement is currently awarded to the Center to Protect Workers' Rights (CPWR). In 2008 the CPWR changed its name to "CPWR: The Center for Construction Research and Training."

EXTERNAL FACTORS

As outlined in the Framework Document, the committee identified *external factors*—defined as actions or forces beyond the Construction Research Program's control—with an important bearing on the ability of the program to effect change (i.e., to have an impact on intermediate and end outcomes) in the workplace.

A fundamental external factor is the fact that the Construction Research Program and NIOSH are research entities lacking rule-making authority. As such, the program can produce knowledge about safety and health hazards in the construction workplace and provide for the transfer of this knowledge through a range of activities. However, the responsibility for issuing and enforcing workplace standards lies with rule-making authorities such as OSHA, and the responsibility for adopting evidence-based best practices lies with construction project owners and contractors. Thus, although the Construction Research Program can make recommendations to these groups and individuals, how its recommendations are used, if at all, is beyond its control.

A second external factor is the segmentation of the industry, which has four distinctly different sectors—residential, commercial, industrial, and heavy construction. This segmentation affects the transfer into practice of research findings and promising interventions. For example, it is easier to translate research into practice in the heavy-construction and industrial sectors, where project owners are more involved and cognizant of health and safety issues, and through labor unions, which have structures for training and information dissemination. Economic factors that some construction owners and contractors may believe to be unfavorable to the implementation of health and safety programs may also play a role.

Other significant factors that have influenced the extent of the program's impact include (1) inadequate funding for conducting the full range of research required and for developing the products, tools, training, and other methods to translate that research into practice; (2) the lack (until recently) of a full-time senior-level person to coordinate the array of projects and activities carried out by the program; and (3) the lack (until 2006) of a Construction Program Manager to advocate for construction research and resources and to hold the program accountable for meeting its objectives.

EVALUATION PROCESS

To ensure an in-depth review of the available information, the committee formed four teams, each one reviewing the part of the Construction Research Program's evidence package corresponding to one of the four research goals and related sub-goals. The teams assessed the various activities of the program and reviewed the body of work resulting from these activities. The committee also assessed the intermediate and end outcomes resulting from these activities.

Using the Framework Document scoring criteria (included here as Box S.1), each committee member provided an independent rating for the relevance and impact of the goal area reviewed by his or her team. The full committee then held substantive discussions to eventually arrive at a consensus on final ratings for the relevance and impact of the total program. The committee also considered external factors that have affected and continue to affect the program.

Evaluation of Relevance: Score of 5

The scoring criteria for relevance are tied to the priority of the research areas focused on and to the level of activity for transferring research to practice. The committee found that the program's priorities for safety-related goals were closely aligned with national and state surveillance data identifying the leading causes of fatalities and injuries. Similarly, its focus on Hispanic workers, the largest "sub-population" within construction, was appropriate and of high priority.

The program's process for prioritizing research on health hazards was not as transparent. In part, this can be attributed to the lack of national and state surveillance data regarding the extent of health hazards caused by specific agents and in comparison with the health hazards caused by other agents. However, it was clear that the research areas chosen for health-related hazards did affect large numbers of workers across the entire construction industry.

The committee also discussed each of the four program goals and their associated research activities at length. The discussion involved a very deliberate process of examining the language for the scoring criteria. Immediate consensus emerged that at a minimum a rating of 4 would apply—that is, research is in priority areas and the program is engaged in appropriate transfer activities. The discussion then turned to a closer examination of the criteria for a score of 5 to determine if the research conducted for each goal was in "high-priority" subject areas and whether the program was "significantly engaged in appropriate transfer activities for completed research projects/reported research results." (The scoring guidance does not allow for the assignment of scores using decimals, e.g., 4.5.)

The committee determined that the Construction Research Program was clearly engaged in high-priority activities given its focus on the leading causes of fatalities (Goal 1), health hazards that affect large numbers of construction workers (Goals 2 and 3), and significant subpopulations (Sub-goal 4.2).

The committee also determined that the Construction Research Program was significantly engaged in appropriate transfer activities. Across the program,

BOX S.1
Framework Document Scoring Criteria for Relevance and Impact

Scoring Criteria for Relevance

5 = Research is in high-priority subject areas and NIOSH is significantly engaged in appropriate transfer activities for completed research projects/reported research results.

4 = Research is in priority subject areas and NIOSH is engaged in appropriate transfer activities for completed research projects/reported research results.

3 = Research is in high priority or priority subject areas, but NIOSH is not engaged in appropriate transfer activities; or research focuses on lesser priorities but NIOSH is engaged in appropriate transfer activities.

2 = Research program is focused on lesser priorities and NIOSH is not engaged in or planning some appropriate transfer activities.

1 = Research program is not focused on priorities and NIOSH is not engaged in transfer activities.

Scoring Criteria for Impact

5 = Research program has made major contribution(s) to worker health and safety on the basis of end outcomes or well-accepted intermediate outcomes.

4 = Research program has made some contributions to end outcomes or well-accepted intermediate outcomes.

3 = Research program activities are ongoing and outputs are produced that are likely to result in improvements in worker health and safety (with explanation of why not rated higher). Well accepted outcomes have not been recorded.

2 = Research program activities are ongoing and outputs are produced that may result in new knowledge or technology, but only limited application is expected. Well accepted outcomes have not been recorded.

1 = Research activities and outputs do not result in or are NOT likely to have any application.

SOURCE: Reprinted from Boxes 2 and 3 of "Framework for the Review of Research Programs of the National Institute for Occupational Safety and Health," reproduced as Appendix A in this report.

research-to-practice (R2P) activities have involved a wide range of industry stakeholders, technologies, training methods, and information-dissemination activities. The program has contributed to the development of OSHA standards and worked directly with state agencies and industry stakeholders to transfer information and protective measures to the worker in the field. Stakeholder groups indicated to the committee that program-generated publications brought value to the industry by offering a means for informing their management, staffs, and members about newly developed or improved industry practices.

On the basis of its determination that the research conducted was high priority in nature and that the program was significantly engaged in appropriate transfer activities, the committee assigned the Construction Research Program a score of 5 for relevance.

The committee did not view this score as a statement that the program could not be improved, however. The high score instead reflects the guidance for ranking established in the Framework Document and the committee's recognition of the financial constraints within which the program has operated. For these reasons, the committee's evaluation is retrospective. During the course of its discussions and evaluation, the committee identified areas on which the program should focus in the future. Those recommendations are prospective and are meant to help ensure that the program continues to work in high-priority areas.

Evaluation of Impact: Score of 4

The committee evaluated the impact of the Construction Research Program using the same process that it used for relevance. The scoring criteria for impact are linked to a program's contributions to worker health and safety based on end outcomes or well-accepted intermediate outcomes. In terms of end outcomes, the committee concluded that the program, through its development of some technologies such as fall-protection equipment and proximity-warning systems, has made some contributions to the overall declines in fatalities and injuries, although the full extent of that impact is not known. Additionally, the program has had a positive impact on the health of workers exposed to asphalt fumes generated during road-paving operations.

The program has also been responsible for a large range of intermediate outcomes. Its research on musculoskeletal disorders is cited in about half of all publications on this topic (NIOSH, 2007). The program has provided evidence for the development of OSHA standards on ergonomics, hearing conservation, respiratory crystalline silica, trenching practices, and lead in construction. Some of these standards have been issued; others have not. However, whether the standards are issued and enforced is beyond the control of the program. Its training and training dissemination activities have been extensive, and it is likely that they have contributed to the prevention and reduction of health and safety hazards at some construction worksites.

The committee also determined that the segmentation of the industry and the less-than-adequate level of resources have had a bearing on the program's impact. Thus, although program-generated publications, technologies, and training are relevant for all segments of the construction industry, their diffusion has varied by construction sector. It is particularly difficult to reach the residential sector because so many residential contractors are self-employed or employ fewer than 10 workers. The level of funding available limits the ability of the program to conduct surveillance research and to provide more direct training to owners and workers in this sector.

Using the scoring criteria for rating the program's impact, the committee determined that the Construction Research Program has made some contributions to construction health and safety as measured by either end outcomes or well-accepted intermediate outcomes. However, committee members had divergent views as to whether these contributions could be classified as major contributions across the entire program. On that basis, the committee assigned the program an impact score of 4. As it did with respect to the program's relevance, the committee made recommendations intended to improve the program's impact in the future.

FUTURE RESEARCH AREAS

The committee was asked to assess the Construction Research Program's effectiveness in targeting new research areas and identifying emerging issues most relevant to future improvements in the protection of workers in the workplace. The committee's analysis of emerging issues and its recommendations for future research are based on discussions with stakeholders, on the NIOSH evidence package, and on individual committee members' knowledge of the construction industry and their backgrounds and expertise. In addition, the committee reviewed a number of priority topics identified by the NIOSH Construction Steering Committee in 2002 as areas where research would be most likely to improve the program's impact. The CSC's topics were grouped into three categories:

1. Health and injury outcome topics that target the following:
    - Leading types of fatal and nonfatal traumatic injuries experienced in construction;
    - Low-back injuries and other cumulative work-related musculoskeletal disorders among construction workers; and

    - Occupational illness topics that focus on respiratory disease and hearing loss. Respiratory disease includes airways disease, asthma, chronic obstructive lung disease, and silicosis.
2. Chemical and physical exposure topics that target the following:
    - Vibration,
    - Asphalt fumes, and
    - Lead and dust particles.
3. Approach and sector topics that target the following groups and issues within construction:
    - Small and self-employed contractors;
    - Special subpopulations at risk within construction, such as Hispanic workers, day laborers, young workers, and aging workers;
    - The role of project design as a primary prevention tool for addressing construction hazards;
    - Addressing work organization in construction and improving the understanding of how it affects health and safety;
    - Working with building owners and clients to promote and evaluate construction best practices; and
    - Leveraging promising approaches from related high-risk sectors such as agriculture and mining into construction.

Some of these topics were further developed into NIOSH Construction Research Program strategic goals in 2005. NIOSH shared these strategic goals as input to the NORA Construction Sector Council in 2006, and most but not all of these 2002 and 2005 topics were subsequently incorporated in some form into the NORA2 Preliminary Draft National Construction Agenda Strategic Goals (Box S.2).

Goals 1.0 through 7.0 of that set are classified as "outcome" goals that will result in actual reductions in injuries, exposures, illnesses, and disorders among construction workers. Goals 8.0 through 14.0 are classified as "contributing-factor" goals. These are defined as factors that represent important influences affecting the likelihood that prevention and control measures and actions are taken on a construction site.

Some of these topics represent areas in which the Construction Research Program is already engaged, offering important opportunities to move research into practice. The committee's recommendations regarding future research areas are presented within the context of the NORA2 goals in Chapter 4. The committee is particularly interested in emphasizing to the Construction Research Program staff that they allocae an increased amount of research time, effort, and resources to the contributing-factors goals within NORA2, that is, Strategic Goals 8.0 through 14.0.

BOX S.2
NORA2 Preliminary Draft National Construction Agenda Strategic Goals

STRATEGIC GOAL 1.0—Reduce construction worker fatalities and serious injuries caused by falls to a lower level.

STRATEGIC GOAL 2.0—Reduce fatal and nonfatal injuries from contact with electricity among construction workers.

STRATEGIC GOAL 3.0—Reduce fatal and serious injuries associated with struck-by incidents associated with objects, vehicles, and collapsing materials and structures.

STRATEGIC GOAL 4.0—Reduce hearing loss among construction workers by increased use of noise reduction solutions, practices, and hearing conservation programs by the construction community.

STRATEGIC GOAL 5.0—Reduce silica exposures and future silicosis risks among construction workers by increasing the availability and use of silica dust controls and practices for tasks associated with important exposures.

STRATEGIC GOAL 6.0—Reduce welding fume exposures and future related health risks among construction workers by increasing the availability and use of welding fume controls and practices for welding tasks.

STRATEGIC GOAL 7.0—Reduce the incidence and severity of work-related musculoskeletal disorders among construction workers in the U.S.

STRATEGIC GOAL 8.0—Increase understanding of factors that comprise both positive and negative construction safety and health cultures; and, expand the availability and use of effective interventions to maintain safe work practices 100% of the time in the construction industry.

For all of the goals, the committee also recommends that the program keep the worker and contractor in mind as the ultimate intended beneficiaries of its R2P efforts. The following are two critical research questions that should remain at the forefront of these efforts: (1) How can the program get vital information to the worker "in the trench" or "on the steel"? and (2) How does the program persuade contractors and workers to use effectively the interventions that are developed through research?

STRATEGIC GOAL 9.0—Improve the effectiveness of safety and health management programs in construction and increase their use in the industry.

STRATEGIC GOAL 10.0—Improve understanding of how construction industry organization factors relate to injury and illness outcomes; and increase the sharing and use of industry-wide practices, policies, and partnerships that improve safety and health performance.

STRATEGIC GOAL 11.0—Increase the recognition and awareness of construction hazards and the means for controlling them through broad dissemination of quality training for construction workers, including non-English speaking workers.

STRATEGIC GOAL 12.0—Increase understanding of how vulnerable worker groups experience disproportionate risks in construction work and expand the availability and use of effective interventions to reduce injuries and illnesses among these groups.

STRATEGIC GOAL 13.0—Increase the use of "prevention through design (PtD)" approaches to prevent or reduce safety and health hazards in construction.

STRATEGIC GOAL 14.0—Improve surveillance at the Federal, State, and private level to support the identification of hazards and associated illnesses and injuries; the evaluation of intervention and organizational program effectiveness; and the identification of emerging health and safety priorities in construction.

SOURCE: Reprinted from NORA Construction Sector Council (2008).

OVERARCHING RECOMMENDATIONS FOR PROGRAM IMPROVEMENT

The overarching recommendations presented and discussed in the subsections below and listed together in Box S.3 apply to the Construction Research Program as a whole. Recommendations regarding specific research topics are presented in Chapter 4, in the section entitled "Analysis of and Recommendations Regarding Emerging Issues and New Research Areas."

BOX S.3
Six Overarching Recommendations for the Construction Research Program

The recommendations listed below apply to the Construction Research Program as a whole and are discussed in the Summary and in Chapter 4. Recommendations regarding specific research topics are presented in Chapter 4, in the section entitled "Analysis of and Recommendations Regarding Emerging Issues and New Research Areas."

The committee recommends that:

1. Research-to-practice (R2P) efforts should involve individuals with the training or with the experience and skills to create strategic diffusion and social marketing plans for National Institute for Occupational Safety and Health research and to evaluate such plans' effectiveness.
2. Consideration should be given to having the majority of research-to-practice efforts of the Construction Research Program conducted through the National Construction Center.
3. High-level attention should be given to determining how to provide program resources that are commensurate with a more robust pursuit of the Construction Research Program's goals.
4. The Construction Program Coordinator and the Construction Program Manager should both be devoted full-time to the Construction Research Program.
5. The National Construction Center should continue to be used as an important component in the Construction Research Program.
6. The Construction Research Program should establish a closer connection with the Occupational Safety and Health Administration and other regulatory or consensus standards organizations to help ensure that the program's research is applied effectively in rule-making efforts.

Transferring Research to Practice

Recommendation 1: Research-to-practice (R2P) efforts should involve individuals with the training or with the experience and skills to create strategic diffusion and social marketing plans for National Institute for Occupational Safety and Health research and to evaluate such plans' effectiveness.

Recommendation 2: Consideration should be given to having the majority of research-to-practice efforts of the Construction Research Program conducted through the National Construction Center.

A number of barriers currently exist within the program structure that limit the R2P efforts and likely their effectiveness. First, although the most recent cooperative agreement for the National Construction Center includes language stipulating that 20 percent of direct costs be used to increase the knowledge base for the effective diffusion of research to practice for construction, in most cases this is not enough to implement more active dissemination strategies and evaluate their effectiveness. Outputs generated by external grantees and partners need to be included in the program's R2P efforts as well. Internally, program researchers have been encouraged to translate research findings to lay publications for target audiences and stakeholders. Indeed, an R2P plan is now a requirement for all internally funded projects.

The individual expertise called for by the committee in Recommendation 1 above does not necessarily need to reside in NIOSH staff. It could also be expertise within other government agencies such as the Centers for Disease Control and Prevention or OSHA or within the private sector, called on by the program to accomplish more effectively its goals for diffusion of research.

Given that NIOSH is a federal agency, the document review process can be lengthy, and limitations are sometimes placed on what can and cannot be said, given that recommendations may be interpreted as policy. The National Construction Center is not constrained by these barriers, however, and thus consideration should be given to having the majority of R2P efforts conducted through the NCC; see Recommendation 2, above. This would allow the program staff to partner with NCC researchers and stakeholders and to focus on conducting the diffusion-related research necessary to determine the optimum ways to reach target audiences.

Resources

Recommendation 3: High-level attention should be given to determining how to provide program resources that are commensurate with a more robust pursuit of the Construction Research Program's goals.

Recommendation 4: The Construction Program Coordinator and the Construction Program Manager should both be devoted full-time to the Construction Research Program.

Recommendation 5: The National Construction Center should continue to be used as an important component in the Construction Research Program.

During its review, the committee concluded that in spite of budget constraints, the Construction Research Program has made an impact on one of the most dangerous and largest of U.S. industries. The total budget for the program from FY 1997 through FY 2007 has, in fact, stayed even or declined slightly in real dollars. It has also been declining as a portion of the total NIOSH budget during all of the period reviewed, 1996-2005. The committee finds the funding level inadequate and recommends that high-level attention be given to determining how to provide program resources that are commensurate with a more robust pursuit of the program's goals (see Recommendation 3, above).

A related matter, addressed in Recommendation 4, above, is that until very recently, NIOSH senior management had not made the commitment to assign at least one full-time senior-level staff person to coordinate the array of projects and activities carried out within the program. The committee supports NIOSH's action in making this a full-time position and recommends that NIOSH continue this practice into the future.

Until 2005, program activities were directed through the CSC, which is composed of representatives from NIOSH divisions and laboratories. In 2006, NIOSH appointed a senior lead team representative as the Construction Program Manager. The committee supports this action and recommends that this position also be devoted full-time to the Construction Research Program. The committee encourages NIOSH to ensure that this position has some level of budgetary authority and management responsibility so that the Construction Program Manager can provide strategic and programmatic leadership and also assist in holding the program accountable for achieving its future research goals.

As indicated in Recommendation 5, above, the committee also recommends that the National Construction Center continue to be used as an important component in NIOSH's Construction Research Program.

Increased Communication with Rule-Making Authorities

Recommendation 6: The Construction Research Program should establish a closer connection with the Occupational Safety and Health Administration and other regulatory or consensus standards organizations to help ensure that the program's research is applied effectively in rule-making efforts.

The committee recommends that the program increase its current level of communication with OSHA and other regulatory or consensus standards organizations about the evidence generated from its research activities. In addition to discussing research findings, program staff should communicate more fully about the economics of occupational disorders and illnesses and their impact on workers and contractors in the industry. Such information will provide valuable supporting documentation for recommendations made by the program with respect to regulatory action. Any role that the Construction Research Program can play in developing or strengthening standards that address risk exposure will likely increase the program's impact on risk reduction for occupational disorders and illnesses.

REFERENCES

CFR (Code of Federal Regulations). 2008. Title 29, Part 671, Sections 20, 21, and 22.

CPWR (Center to Protect Workers' Rights). 2007. The Construction Chart Book: The U.S. Construction Industry and Its Workers. 4th ed. Washington, D.C.: Center for Construction Research and Training.

NIOSH (National Institute for Occupational Safety and Health). 2007. NIOSH Construction Research Program Evidence Package. Washington, D.C., July.

NORA (National Occupational Research Agenda) Construction Sector Council. 2008. National Occupational Research Agenda (NORA) Draft Preliminary Comment Version. Available at http://www.cdc.gov/niosh/nora/comment/public/constdraftdec2007/pdfs/constdraftdec2007.pdf. Accessed August 15, 2008.

Wiatrowski, W.J. 2004. Occupational injury and illness: New recordkeeping requirements. Monthly Labor Review Online. December, pp. 10-24. Available at http://www.bls.gov/opub/mlr/2004/12/art2abs.htm. Accessed July 15, 2008.

1

Introduction

Construction is a large, dynamic, and complex industry that plays an important role in the U.S. economy and the global economy. Construction work ranges from major civil engineering and infrastructure projects (dams, highways, airports) involving a multitude of individual construction firms, to the construction and renovation of residential, commercial, and industrial structures, to routine residential repairs. In 2005, the industry accounted for $1.2 trillion of construction put in place, the equivalent of 13 percent of the U.S. gross domestic product, and employed about 11 million workers (U.S. Census Bureau, 2005). The Business Roundtable has called construction a "seminal" industry because the price of every factory, office building, hotel, or power plant that is built affects the prices that must be charged for the goods and services produced in it. These prices affect U.S. consumers and the ability of U.S. businesses to compete in a global market (BRT, 1983).

Construction is also one of the most dangerous industries for workers. Among all industries, construction had the fourth highest rate of fatalities (following agriculture, mining, and transportation) and the second highest rate of nonfatal injuries and illnesses (after transportation) in 2005 (BLS, 2006a,b).

Hazards in construction work include the following: working at heights, in excavations and tunnels, on highways, and in confined spaces; exposure to high levels of noise, chemicals, and high-voltage electric lines; the use of power tools and heavy equipment; manual materials handling; and sustained awkward postures. The leading causes of death among construction workers are falls from elevations, being struck by vehicles and equipment, electrocution, machine-related incidents, and being struck by objects (NIOSH, 2007). Significant health risks include hearing loss, silicosis, musculoskeletal disorders, skin diseases, and health effects associated with exposures to lead, asphalt fumes, and welding fumes. Additional health hazards and associated diseases include fume fever (metal, polymer), cadmium poisoning, carbon monoxide poisoning, acute inhalation injury (nitrogen dioxide, ozone, phosgene), manganese poisoning,[1] asbestosis,[2] acute solvent syndrome,[3] peripheral neuropathy,[4] allergic contact dermatitis,[5] chronic obstructive pulmonary disease,[6] occupational asthma,[7] and hypersensitivity pneumonitis.[8]

The National Institute for Occupational Safety and Health (NIOSH) was established by the Occupational Safety and Health Act of 1970 (Public Law 91-596) to "conduct (directly or by grants or contracts) research, experiments, and demonstrations relating to occupational safety and health, including studies of psychological factors involved, and relating to innovative methods, techniques, and approaches for dealing with occupational safety and health problems." As a research agency, NIOSH does not have the authority to establish and enforce regulations on workforce safety and health. Regulatory and enforcement authority rests with other organizations, including the Occupational Safety and Health Administration (OSHA), which was created by the same Public Law 91-596 that established NIOSH. Organizationally, NIOSH is part of the Centers for Disease Control and Prevention (CDC) in the U.S. Department of Health and Human Services, while OSHA is an agency in the U.S. Department of Labor.

NIOSH has conducted construction-related research activities since the 1970s, including research on silicosis and on exposures to lead, asphalt fumes, and welding fumes. The program was formalized in 1990 when Congress allocated specific funds with which it directed NIOSH to develop a comprehensive prevention program focused on health and safety problems affecting construction workers by expanding existing NIOSH activities in the areas of surveillance, research, and intervention.

---

[1]Manganese poisoning is a toxic condition resulting from chronic exposure to manganese, usually as the result of lead or arc welding.

[2]Asbestosis is a respiratory disease caused by inhaling asbestos fibers.

[3]Acute solvent syndrome results from acute exposure to toxic cleaners, degreasers, and solvents.

[4]Peripheral neuropathy is nerve damage caused by trauma from external agents.

[5]Allergic contact dermatitis is an inflammation of the skin caused by direct contact with an irritating substance.

[6]Chronic obstructive pulmonary disease (COPD) is caused by prolonged exposure to fumes, dust, or fibers resulting in lung damage.

[7]Occupational asthma is a lung disorder in which various substances found in the workplace lead to breathing difficulties.

[8]Hypersensitivity pneumonitis is an inflammation of the lungs due to breathing in a foreign substance, usually certain types of dust, fungus, or molds.

NIOSH is authorized to carry out the following:

- Develop recommendations for occupational safety and health standards,
- Conduct research on worker safety and health,
- Conduct training and employee education,
- Develop information on safe levels of exposure to toxic materials and harmful physical agents and substances,
- Conduct research on new safety and health problems,
- Conduct on-site investigations (health hazard evaluations) to determine the toxicity of materials used in workplaces, and
- Fund research by other agencies or private organizations through grants, contracts, and other arrangements (CFR, 2008).

STUDY CHARGE AND EVALUATION COMMITTEE

In 2004, NIOSH requested that the National Academies conduct evaluation reviews of up to 15 specific NIOSH research programs, including the Construction Research Program, to assess the relevance and impact of its research in reducing workplace injury and illness. The statement of task for the overall NIOSH effort stated the following:

> Each evaluation will be conducted by an ad hoc committee, using a methodology and framework developed by the Committee to Review NIOSH Research Programs (Framework Committee).
>
> Each evaluation committee will review the program's impact, relevance, and future directions. The evaluation committee will evaluate not only what the NIOSH research program is producing, but will also determine whether it is appropriate to credit NIOSH research with changes in workplace practices, hazardous exposures, and/or occupational illnesses and injuries, or whether the changes are the result of other factors unrelated to NIOSH.
>
> The program reviews should focus on evaluating the program's impact and relevance to health and safety issues in the workplace and make recommendations for improvement. In conducting the review, the evaluation committee will address the following elements:
>
> 1. Assessment of the program's contribution through occupational safety and health research to reductions in workplace hazardous exposures, illnesses, or injuries through:
>
> a. an assessment of the relevance of the program's activities to the improvement of occupational safety and health, and
>
> b. an evaluation of the impact that the program's research has had in reducing work-related hazardous exposures, illnesses, and injuries.

> The evaluation committee will rate the performance of the program for its relevance and impact using an integer score of 1 to 5. Impact may be assessed directly (e.g., reductions in illnesses or injuries) or, as necessary, using intermediate outcomes to estimate impact. Qualitative narrative evaluations should be included to explain the numerical ratings.
>
> 2. Assessment of the program's effectiveness in targeting new research areas and identifying emerging issues in occupational safety and health most relevant to future improvements in workplace protection. The committee will provide a qualitative narrative assessment of the program's efforts and suggestions about emerging issues that the program should be prepared to address.

The Framework Committee developed a document (Appendix A) to guide each research program evaluation committee as it reviews materials provided by various NIOSH programs and to provide the rationale for determining final scores for each program's impact and relevance.

The Committee to Review the NIOSH Construction Research Program, which authored this report, was structured to include members with expertise in medicine, construction research, performance measurement and management, construction safety, engineering, economics, epidemiology, industrial hygiene, and control technology. Committee members have worked in academia, industry, government, and labor unions (Appendix D).

To conduct its evaluation, the committee met three times between July and December 2007 and corresponded through e-mail and conference calls between and after the meetings. The committee's review was based in large part on material submitted by NIOSH in the form of an evidence package (NIOSH, 2007) containing more than 500 descriptive pages of Construction Research Program goals, activities, products, and impacts. Staff from NIOSH and the Construction Research Program made 18 presentations to the committee during open-session meetings (Appendix B). The committee also heard from 17 stakeholder organizations during an open-session meeting, including representatives from the OSHA Directorate of Construction, labor unions, state governments, insurance companies, worker and labor management organizations, organizations of occupational safety and health experts, and other construction-relevant professional organizations. Throughout the study, additional information was received from the Construction Research Program staff in response to requests from committee members.

The time period chosen by the committee for the evaluation was 1996 through 2005. To the extent possible and practical, the statistics and other information cited in this report correspond to that time period.

TRENDS IN CONSTRUCTION WORKPLACE SAFETY AND HEALTH

Fatality rates and nonfatal injury and illness rates provide a benchmark for measuring the success of an industry's occupational safety and health strategies. Between 1992 and 2005, 16,000 construction workers died from work-related injuries. In 2005 alone, 1,243 construction workers died from job-related traumatic injuries. This number accounted for 22 percent of job-related deaths across all industries, a figure that is disproportionately high given that construction workers account for 8 percent of the workforce. The death rate for construction workers was almost three times that of full-time workers for all industries: 11.1 deaths per 100,000 construction workers compared with 4.2 deaths per 100,000 workers in all industries. Among all sectors, construction had the fourth highest rate of fatalities in 2005 (after agriculture, mining, and transportation) (CPWR, 2007, Section 32).

Nonetheless, when compared with statistics from the past, these numbers indicate significant improvements in construction workplace safety. Between 1992 and 2005, the overall rate of construction-related fatalities declined from 14.3 to 11.1 per 100,000 workers, which translates to 350 fewer deaths per year for a workforce of 11 million people.

Nonfatal injuries and illnesses are also an important indicator of workplace safety and health. In 2005, construction workers accounted for 414,900 injuries and illnesses, a rate of 239.5 per 10,000 workers, the second highest of all industries. Of the 414,900 cases, 157,100 were serious enough to result in lost workdays (CPWR, 2007). As with fatalities, the rate of nonfatal injuries and illnesses also declined significantly, although the exact numbers are not clear owing to changes in reporting requirements.[9]

The statistics for nonfatal injuries and illnesses are misleading in one sense, however: Of the total cases reported, only 2.5 percent were illness-related (CPWR, 2007). Thus, for construction, this statistic is in actuality indicative of nonfatal injuries. Also, the data fail to include the large portion of the construction workforce that is self-employed or employed by firms with fewer than 10 workers.

[9]In 2002, the Occupational Safety and Health Administration implemented a number of changes in the definitions of injury and illness cases recorded by employers. The new definitions in turn resulted in changes in occupational injury and illness statistics provided by the Bureau of Labor Statistics. There is some disagreement as to the overall effect of these changes. For example, one author writes that "while these data follow the trend of declining cases and rates seen throughout the past decade, because of the change in definition they cannot be compared with data from prior years" (Wiatrowski, 2004). In contrast, others note that although the changes in coding systems have significantly affected the compatibility of injury and illness data for construction subsectors over time, the impact on the construction industry as a whole is relatively small (CPWR, 2007, Section 32).

Data on construction-related illnesses are, in fact, difficult to come by. National statistics are not available to guide all occupational disease research. National estimates do indicate that occupational illness is an important problem and that the total burden from the number of deaths due to occupational illness is likely to exceed that for occupational injury by a factor ranging from 4 to 11 (Steenland et al., 2003). A number of studies with data from death certificates are also available; the studies indicate that occupational diseases remain significantly undercounted, and thus it can be deduced that deaths associated with occupational disease are underestimated. For example, construction was the most frequently listed industry on asbestosis and silicosis death certificates from 1990 to 1999 (24.6 percent and 13.4 percent, respectively) (CDC, 2008). Lacking specific data related to illnesses, researchers must rely on the picture provided by a variety of sources such as national data, state-level illness statistics, knowledge and extrapolation of construction exposures, and international surveillance findings.

To set the context for evaluating the Construction Research Program, it is also important to describe some characteristics of the construction industry and its workforce that directly influence the relevance of research and the potential impact of research-based activities and products.

CHARACTERISTICS OF THE CONSTRUCTION INDUSTRY

National Statistics for the Industry

National statistics for the construction industry are limited in their usefulness because they do not include the almost 2.5 million self-employed "one person" businesses without other paid employees, or approximately 1.5 million public employees performing construction (U.S. Census Bureau, 2005). The data available from the Bureau of Labor Statistics divide construction into three subsectors using the North American Industry Classification System (NAICS) (Table 1.1).

The Construction of Buildings subsector comprises establishments involved in constructing residential, industrial, commercial, and institutional buildings. The Heavy and Civil Engineering Construction subsector includes establishments involved in infrastructure projects—for example, water, sewer, oil, and gas pipelines; roads and bridges; and power lines. The Specialty Trade Contractors subsector engages in activities such as plumbing, electrical work, masonry, carpentry, and roofing that are generally needed in the construction of all building types. Thus, while two of the subsectors refer to types of construction projects, the third refers to types of workers who work on all types of projects. This categorization obscures differences in injury and illness rates among different types of projects.

TABLE 1.1 Construction Industry Subsectors as Defined by the North American Industry Classification System

NAICS Code	Construction Subsector	Employment 2005[a]	Nonfatal Rates, 2004[b]	
			Injury	Illness
236	Construction of Buildings	1,782,200	5.5	13.4
237	Heavy and Civil Engineering Construction	974,800	5.8	16.2
238	Specialty Trade Contractors	4,714,000	6.6	13.4
23	Construction Sector	7,416,000	6.2	13.8

[a]Excludes self-employed and publicly employed workers.
[b]Injury rates per 100 employees per year; illness rates per 10,000 employees per year.
SOURCE: CPWR (2007).

Construction workers across the entire industry are typically younger than the national labor force, although the average age in the industry has been increasing. The majority of construction workers are ages 25 to 54 (75.4 percent), are male (90.3 percent), and are white (90.8 percent) (NIOSH, 2007). However, these numbers mask significant differences within segments of the industry, such as the proportions of unionized and non-union workers, the percentages of workers in skilled trades, and those in unskilled or manual labor jobs.

Hispanics are the fastest-growing ethnic group and the largest minority group in the United States, accounting for 14 percent of the U.S. population. This growing population works in some of the most dangerous industries in the nation, with construction having a larger share of Hispanic workers than any U.S. industry except agriculture. The number of Hispanic workers in construction tripled in the past decade to 2.6 million in 2005 (or 23 percent of all construction workers). Almost 75 percent of Hispanic construction workers were born outside the United States and about two-thirds (or 1.7 million) were not U.S. citizens in 2005 (CPWR, 2007).

The construction industry is overwhelmingly one of small establishments. Of the 710,000 construction firms with payrolls in the United States in 2002, almost 80 percent had fewer than 10 employees, accounting for 24 percent of the construction workforce. In contrast, only 585 construction firms (less than 1 percent) had 500 or more employees (8 percent of construction workers). Ninety-eight percent of all firms had fewer than 100 workers (79 percent of the construction workforce), while 2 percent of all firms had 100 or more workers (21 percent) (CPWR, 2007).

This type of statistical categorization is important in determining the rates and causes of fatalities, injuries, and illnesses, as well as demographic changes across and within the industry. This information, in turn, is important in determining which areas of research should be of high priority. However, statistical characterizations lose much of the richness and variation that characterize construction work in the field. They also fail to convey the relative ease or difficulty of influencing worker health and safety in different sectors and, in turn, the ability of research-based activities to have an impact on fatalities, injuries, illnesses, or special populations.

To provide context for its evaluation of both the relevance and the impact of the NIOSH Construction Research Program, the committee considered it important not only to include statistical breakdowns that affect relevance but also to describe briefly construction project processes and stakeholders and the segmentation of the industry from the perspective of industry practitioners. This segmentation has implications for the potential impact of the Construction Research Program on construction workplace health and safety.

Projects and Stakeholders

The construction industry delivers buildings of all types, manufacturing and industrial facilities, civil infrastructure, and public works. Each construction project is initiated by an "owner," which may be a government entity, a corporation, or an individual. In addition to the owner, a construction project may involve architects, engineers, general contractors, subcontractors, members of trade unions, skilled trades persons, manual laborers, suppliers, financing institutions, legal representatives, insurance companies, and others. All of these stakeholders operate in an environment in which there is continual demand to deliver projects in less time and at lower cost.

However, not all individuals involved in construction projects identify themselves as working in the construction industry. For example, owners, architects, and engineers serve the industry but may not say that they "work in construction." Similarly, many manufacturers of products used in the construction industry are classified as just that—manufacturers. The boundary is more subtle for companies that fabricate products for use in construction. The involvement of such an array of stakeholders poses two challenges: (1) determining who is responsible and accountable for workplace safety and health and (2) determining who is in the best position to use and implement research-based information, technologies, and products.

Segmentation of the Industry

In contrast to the NAICS categorization of three sectors—Construction of Buildings, Heavy and Civil Engineering Construction, and Specialty Trade Contractors—many industry practitioners consider construction to have at least four distinct sectors—residential, commercial, industrial, and heavy construction.[10] Specialty trade contractors and manual laborers are involved in each of these sectors.

The sectors delineated by industry practitioners differ from each other in terms of the characteristics of project owners and their sophistication and/or involvement in the construction process, the complexity of projects, the source and magnitude of financial capital, required labor skills, the use of specialty equipment and materials, design and engineering processes and knowledge, and other factors. When viewed in this way, radical differences are apparent among the sectors regarding worker training, and owner and contractor surveillance of workplace safety and health. These differences often determine the level of availability of health and safety interventions. For example, workers in unions generally receive training through apprenticeship programs, often including some training related to health and safety. Established labor-management committees in unions provide a distinct structure for transferring training and engineering controls to workers in the field. Such structures do not typically exist for non-union workers in small firms or manual laborers. Similarly, owners that contract for multiple projects (e.g., large corporations with multiple facilities) and large contractors have greater access to and more resources for participating in owner and contractor associations and organizations that identify best practices than do owners who build only one project or firms with fewer than 10 employees.

The residential sector builds single-family houses, townhouses, and low-rise (up to five-story) multifamily apartments and condominiums and accounts for about 35 percent of the total construction value put in place annually. In 2005, construction was started on approximately 1.72 million new single-family homes (CPWR, 2007). However, the total number of establishments and total number of workers involved in residential construction is not known owing to current methods for gathering statistical data (CPWR, 2007). What is clear is that with the exception of a few "national" contractors that may build up to 20,000 units per year (e.g., Pulte Homes), most residential contractors employ fewer than 10 workers. Residential construction workers and subcontractors tend to move among several projects at any one time and may work as subcontractors to several different general contractors. Some specialty trade workers may have formal training, but many workers, including manual laborers, have only on-the-job experience. In an environment of thousands of small firms and transient workers, gathering data through surveillance and other research techniques or disseminating health- and safety-related information and products is difficult. Compliance with regulations

[10]Some practitioners would suggest that transportation-related projects be treated as a fifth segment of construction based on the characteristics of these types of projects (Hinze, 2001).

or the use of best practices (e.g., proper trenching operations) is problematic: Construction contractors may not have the knowledge, training, or incentives to apply best practices, while regulatory enforcement agencies such as OSHA do not have the resources to inspect small projects, but must instead focus on large projects employing greater numbers of workers.

The commercial sector builds projects such as schools, churches, high-rise multifamily buildings, offices, and retail buildings, among others. This sector accounts for about 25 percent of the total construction value put in place per year. Construction firms and contractors working in this sector may have a mix of large and small projects and a larger, more stable group of full-time workers and subcontractors. Some of this sector's workers may belong to labor unions and may have specialized training through apprenticeships. Gathering and disseminating health and safety information in this type of environment is less problematic than doing so in the residential sector: In the commercial sector, project and business managers tend to have more professional training and are more aware of the impacts and costs of injuries and illnesses. In this sector, compliance with health and safety regulations is more likely to occur as a result of management-initiated practices or the occasional OSHA inspection.

The industrial sector delivers projects such as manufacturing plants and oil refineries and accounts for about 25 percent of total construction value put in place annually. The owners of industrial projects, usually large corporations, typically build them to produce the products that they market. Because such projects are specialized, cost hundreds of millions of dollars, and are integral to the business "bottom line," owners are more likely to be closely involved in such projects. Contractor firms working in this sector tend to be large, sophisticated firms, and their workers are likely to be trained and certified, by trade associations, labor unions, or sometimes the contractor. For some types of projects, both owners and contractors are members of professional organizations, such as the Construction Industry Institute, that share best practices. Workers tend to stay on one site, working on one project at a time rather than moving among several sites simultaneously. In this environment, it is much easier to identify those with a direct interest in worker safety and health and then to gather and disseminate research-based information.

The heavy-construction sector delivers large infrastructure projects, including dams; water, sewer, and gas lines; tunnels, highways, and bridges; and airports. Governmental entities serve as the owner of many but not all such projects. Construction firms working in this sector range from relatively small specialized contractors to large, national firms. Much of the work involves the use of heavy equipment and may require fewer workers per project than are needed in other sectors. As with industrial-type projects, the awareness and involvement in health and safety issues by owners and contractors in the heavy-construction sector is relatively high.

REFERENCES

BLS (Bureau of Labor Statistics). 2006a. Fatality Data. Available at http://www.bls.gov/iif/oshcfoi1.htm. Accessed June 15, 2008.

BLS. 2006b. Injury and Illness Data. Available at http://www.bls.gov/iif/oshsum.htm. Accessed June 15, 2008.

BRT (Business Roundtable). 1983. More Construction for the Money: Summary Report of the Construction Industry Cost Effectiveness Project. Available at http://www.ce.berkeley.edu/~tommelein/BRTMORECONSTRUCTIONFORTHEMONEY.pdf. Accessed June 20, 2008.

CDC (Centers for Disease Control and Prevention). 2008. World Report Data. Available at http://www2a.cdc.gov/drds/WorldReportData/pdf/Table03-06.pdf. Accessed June 15, 2008.

CFR (Code of Federal Regulations). 2008. Title 29, Part 671, Sections 20, 21 and 22.

CPWR (Center to Protect Workers' Rights). 2007. The Construction Chart Book: The U.S. Construction Industry and Its Workers. 4th ed. Washington, D.C.: Center for Construction Research and Training.

Hinze, J.W. 2001. Construction Contracts. New York: McGraw-Hill Companies.

NIOSH (National Institute for Occupational Safety and Health). 2007. NIOSH Construction Research Program Evidence Package. Washington, D.C., July.

Steenland, K., C. Burnett, N. Lalich, E. Ward, and J. Hurrell. 2003. Dying for work: The magnitude of U.S. mortality from selected causes of death associated with occupation. American Journal of Industrial Medicine 43:461-482.

U.S. Census Bureau. 2005. Nonemployer Statistics and Construction Spending. Available at http://www.census.gov/epcd/nonemployer/2005/us/US000_23.htm and http://www.census.gov/const/C30/ototsa2005.pdf. Accessed July 15, 2008.

Wiatrowski, W.J. 2004. Occupational injury and illness: New recordkeeping requirements. Monthly Labor Review Online. December, pp. 10-24. Available at http://www.bls.gov/opub/mlr/2004/12/art2abs.htm. Accessed July 15, 2008.

2

The NIOSH Construction Research Program

The mission of the Construction Research Program of the National Institute for Occupational Safety and Health (NIOSH) is "to eliminate occupational diseases, injuries, and fatalities among individuals working in these industries through a focused program of research and prevention" (NIOSH, 2007). The Construction Research Program has evolved over time in response to congressional directives, to several internal initiatives, and to construction-related safety and health issues. Chapter 2 describes the evolution of the program, its strategic planning processes, the program's structure and administration, and its resources. External factors that are beyond the control of the program but that nonetheless affect its relevance and impact in reducing fatalities, injuries, and illnesses on construction worksites are also described.

PROGRAM EVOLUTION

Although the formal Construction Research Program was not created until 1990, construction-relevant research activities at NIOSH started in the 1970s. These activities included large national surveillance and investigation activities, such as the National Occupational Hazard Survey, the National Occupational Exposure Survey, the National Traumatic Occupational Fatalities surveillance program, Fatality Assessment and Control Evaluations, the Sentinel Event Notification Systems for Occupational Risk, and the Adult Blood Lead Epidemiology and Surveillance program. NIOSH staff conducted research that addressed worker exposures to lead, asphalt fumes, and silica. Grants and contracts were awarded to support epidemiological studies investigating health risks associated with the painting trade, silicosis and its association with sandblasting, and safety profiles for specific construction activities. NIOSH researchers were also developing and disseminating criteria documents, providing research results to the Occupational Safety and Health Administration (OSHA) regarding the health and safety aspects of noise, ultraviolet radiation, elevated workstations, crystalline silica, asphalt fumes, construction work in confined spaces, excavation, and occupational exposure to hand-arm vibration for construction workers.

In 1990, following hearings about the level of resources and programs targeting construction safety, Congress directed NIOSH to "develop a comprehensive prevention program directed at health problems affecting construction workers by expanding existing NIOSH activities in areas of surveillance, research, and intervention" (NIOSH, 2007) and allocated funds for NIOSH to do so. Between 1990 and 1994, NIOSH conducted several national conferences on construction safety and health, issued cooperative agreements to encourage extramural research,[1] and established a task group to prepare a plan and budget for construction research.

By 1994, extramural research included state demonstration projects and several cooperative agreements and research grants. In the same year Congress directed NIOSH to establish a new 5-year cooperative agreement with the construction trades to develop a center for prevention-oriented strategies and programs. After NIOSH issued a request for applications (RFA) and held a competition, the agreement for a National Construction Center (NCC) was awarded to the Center to Protect Workers' Rights (CPWR) and a CPWR consortium of 10 academic institutions.[2] The CPWR had been created by the Building and Construction Trades Department of the AFL-CIO in 1990 to conduct applied construction safety and health research, training, and medical screening and to provide other related services.

In 1995, the NIOSH director requested an external review of the Construction Research Program. The review resulted in the establishment of a Construction Steering Committee (CSC), which included a chairperson and a representative from each NIOSH division and laboratory.[3] The mandate of the CSC was to increase internal and external communication between researchers working on

[1] Research conducted by entities and individuals outside of NIOSH.

[2] CPWR was renamed "CPWR: The Center for Construction Research and Training" in 2008.

[3] Applied Research and Technology; Surveillance, Hazard Evaluations, and Field Studies; Education and Information; Respiratory Disease Studies; Safety Research; Health Effects; Pittsburgh Research Laboratory; Spokane Research Laboratory, the Office of Extramural Projects; and National Personal Protective Technology. NIOSH's laboratories are located in Pittsburgh, Pennsylvania; Spokane, Washington; Cincinnati, Ohio; and Morgantown, West Virginia.

construction health and safety projects and to establish a formal review process for construction projects proposed by NIOSH researchers.

In 2001, Congress directed NIOSH to expand the Construction Research Program and appropriated funds for a 3-year grant to Purdue University. This eventually led to the formation of a Construction Safety Alliance partnership project based at Purdue. This grant included the funding of efforts for the NIOSH Student Engineering Team partnership and provided the program with necessary access to expertise, facilities, and equipment to accomplish a broad scope of work. It gave the program exposure on campus, primarily providing single-discipline graduate training in industrial hygiene, occupational health nursing, occupational medicine, and occupational safety.

In 2003, an RFA was issued for the second 5-year NCC Cooperative Agreement announcement. The CPWR scored highest in the competition and was awarded the contract. In separate actions, individual projects submitted by Purdue University and Virginia Tech were also funded.

STRATEGIC PLANNING AND RESEARCH GOALS

As the Construction Research Program has evolved, so have its processes for strategic planning and the development of research goals. The drivers of the research topics and activities in the first years of the program included congressional mandates and directives, construction stakeholder input obtained at national conferences, and the CSC. A plan was developed that included goals related to surveillance, research, and intervention development. It also identified construction workers as particularly vulnerable for fatal and nonfatal injuries, envisioned collaboration with other agencies, and recognized the importance of input from construction industry representatives, both labor and management, on research and intervention needs.

In the early 1990s, NIOSH and its public and private partners used a consensus-building process to set priorities for a 10-year period for all occupational safety and health research, including construction. Unveiled in 1996, the first National Occupational Research Agenda (NORA1) identified 21 research priorities grouped into three categories: disease and injury, work environment and workforce, and research tools and approaches. Throughout NIOSH, a crosscutting approach to meeting the NORA research priorities was established, resulting in the use of a matrix approach to track, manage, and report on NORA-related research for construction and other high-risk sectors. To communicate its strategic planning priorities to its internal and extramural stakeholders, the CSC prepared annual guidance related to research needs and priorities, asking for proposals that focused on NORA priority areas relevant to construction.

The development of RFAs for the second and third cooperative agreements for a National Construction Center also served as opportunities to evaluate program goals and adjust them to meet identified gaps in construction research. The 1999 RFA was structured to maintain and expand ongoing surveillance activities while generating new extramural research on construction interventions, information and technology transfer, and preventive systems research. By 2003, it was recognized that the adoption of research-based solutions by the industry had been uneven and uncertain, owing in part to gaps in the understanding of how to effectively transfer research results into practice. As a consequence, the third NCC RFA stipulated that 20 percent of direct costs were to be directed to research-to-practice (R2P) translation projects and emphasized the need to measure impacts.

CURRENT PROGRAM STRUCTURE AND ADMINISTRATION

NIOSH has formalized its Construction Research Program using a three-component structure: NIOSH-wide intramural research and surveillance programs, a National Construction Center cooperative agreement, and support grants and agreements for investigator-initiated extramural research projects. Figure 2.1 provides a conceptual overview of the NIOSH Construction Research Program components and the research focus of each component.

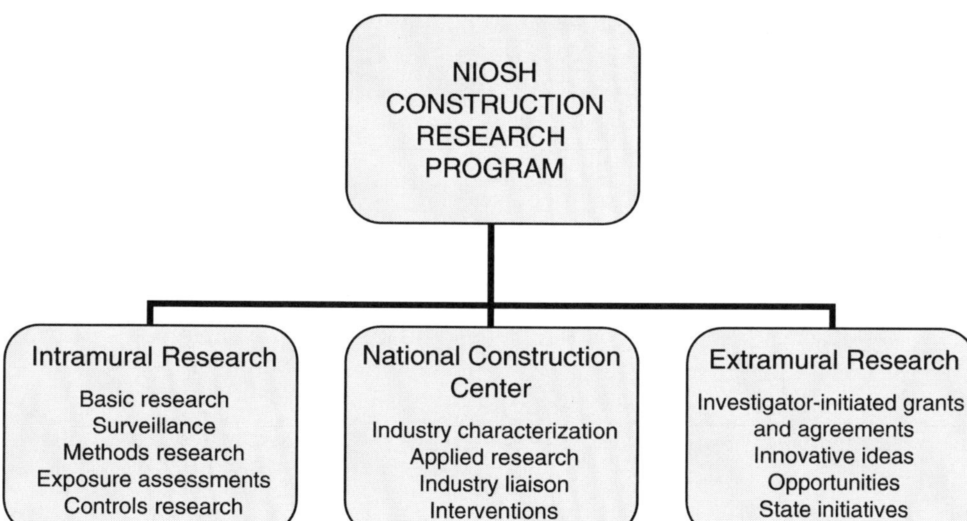

FIGURE 2.1 Components of the NIOSH Construction Research Program.

Intramural research is carried out by NIOSH researchers assigned to several divisions and associated laboratories throughout NIOSH, rather than by a distinct construction research organizational entity. As such, the program operates as a matrix organization within NIOSH. Activities that focus on basic research, surveillance, research methods, exposure assessments, and controls research are managed through a Construction Coordinator and the CSC. The CSC monthly meetings are conducted by videoconference or conference call, and two annual face-to-face meetings are held. NIOSH conducts intramural project reviews at the division level. Each division oversees the review process using established guidelines for seeking external peer reviews, documenting the results, and certifying the review process.

The NCC is operated under a competitively awarded (scientifically and programmatically reviewed) 5-year cooperative agreement. The NCC is currently awarded to CPWR: The Center for Construction Research and Training. The NCC focuses on applied research, creating liaisons with the construction industry, and developing research-based interventions. It is supported through government funding from NIOSH and also from the National Institute for Environmental Health Sciences, the U.S. Department of Energy, U.S. Department of Labor, and U.S. Department of Defense.

Research for NIOSH's Construction Research Program is conducted by NCC staff dedicated to construction research and through the NCC consortium of universities, which includes 26 individual principal investigators affiliated with 19 universities, institutes, and other organizations. The NCC disseminates safety and health-related information through the Internet. The NCC's Library of Construction Occupational Safety and Health (eLCOSH) provides user-friendly information in English, Spanish, and other languages from a wide range of sources that include labor-management programs, trade magazines, universities, and government agencies.[4] The NIOSH Construction Coordinator regularly meets with senior NCC personnel to coordinate research activities.

In addition to the work conducted by the NCC for the Construction Research Program, extramural research is conducted for the program through investigator-initiated extramural grants and cooperative agreements (all scientifically and programmatically reviewed), and through support for state health department investigators working on construction health and safety surveillance and state-level interventions. The program staff also interacts with the Center for Innovation in Construction Safety and Health at Virginia Tech.

Research conducted in any one component of the program is leveraged through interactions with researchers in the other two components. Such interactions

---

[4]See http://www.cdc.gov/elcosh.

occur during regularly scheduled meetings, construction conferences, and other construction-specific networking opportunities.

Program Resources

Total annual funding for the Construction Research Program between fiscal year (FY) 1997 and FY 2007 has averaged about $17.8 million, ranging from a high of $20.3 million in FY 1997 to a low of $13.8 million in FY 2007 (Figure 2.2) (NIOSH, 2007).

When adjusted for inflation and changes in technologies, the funding level for the program has declined in terms of real purchasing power (Figure 2.3). Funding for intramural research and extramural grants, not including the NCC, has fluctuated accordingly.

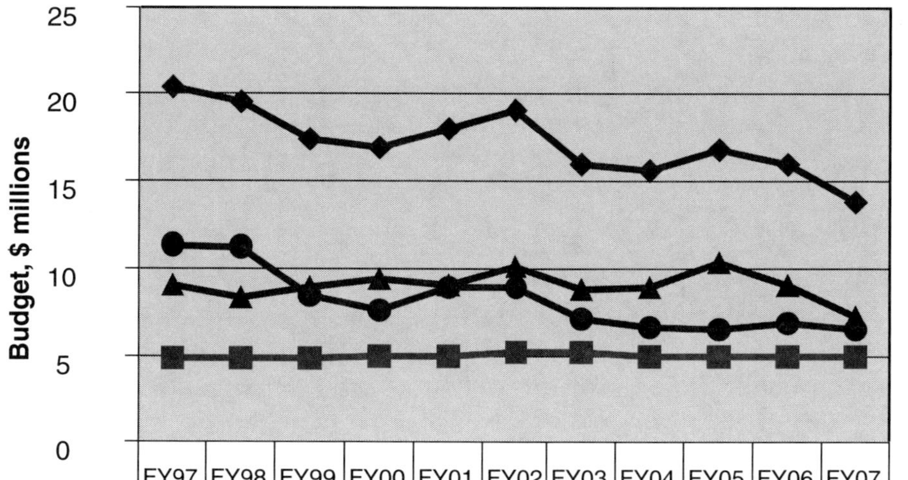

	FY97	FY98	FY99	FY00	FY01	FY02	FY03	FY04	FY05	FY06	FY07
Total Budget	$20.4	$19.6	$17.3	$17.0	$18.0	$19.1	$15.9	$15.6	$16.8	$15.9	$13.8
Intramural	$ 9.1	$ 8.3	$ 8.9	$ 9.4	$ 9.1	$10.2	$ 8.8	$ 9.0	$10.3	$ 9.0	$ 7.3
Extramural	$11.3	$11.3	$ 8.4	$ 7.5	$ 9.0	$ 8.9	$ 7.2	$ 6.7	$ 6.5	$ 6.8	$ 6.5
Center	$ 4.8	$ 4.9	$ 4.9	$ 5.0	$ 5.0	$ 5.2	$ 5.2	$ 5.0	$ 5.0	$ 5.0	$ 5.0

Fiscal Year

FIGURE 2.2 Construction Research Program funding history, FY 1997–FY 2007. NOTE: The costs include staff salaries and overhead. SOURCE: NIOSH (2007).

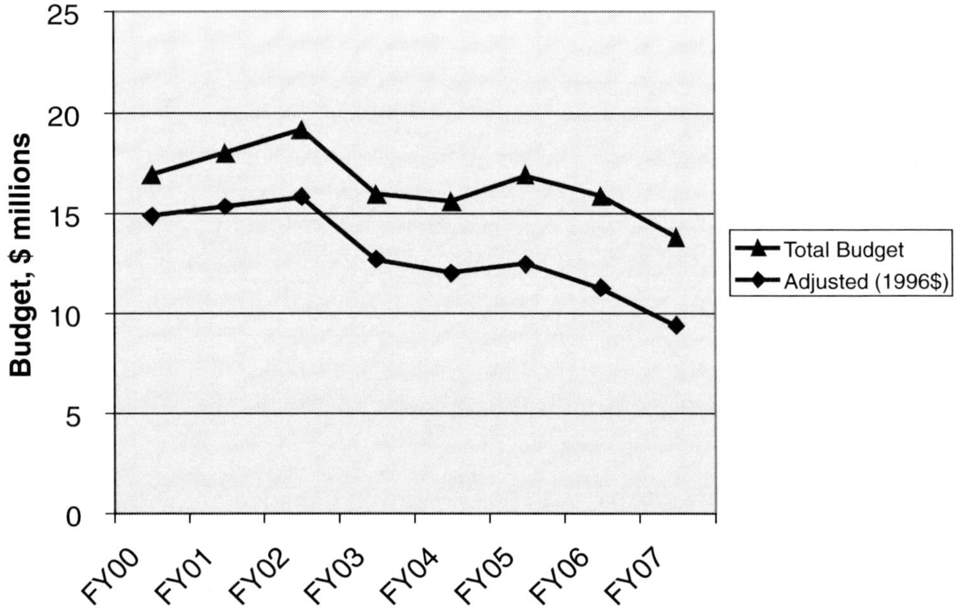

FIGURE 2.3 Construction Research Program funding, FY 2000–FY 2007, adjusted by the Biomedical Research and Development Price Index (BRDPI). NOTE: The BRDPI adjusts for inflation and the increased costs of conducting scientific investigations due to new technologies and other factors. SOURCE: NIOSH (2007).

Because the Construction Research Program is a matrix program, internal staff resources available for intramural research must be aggregated across divisions and reported as full-time equivalents (FTEs). In FY 2007, the research commitment was 56 FTEs, representing individuals from a variety of disciplines including behavioral sciences, epidemiology, safety engineering, safety management, statistics, general engineering, communications, industrial hygiene, and health science. CSC representatives typically allocate 10 percent of their time to program management and direction. The Construction Coordinator, who allocated 25 percent of his time to the program in 2000, currently allocates 100 percent of his time to the program. Beginning in 2006, a senior lead team representative was designated as Construction Program Manager and the CSC representatives transitioned to also representing NIOSH on the NORA Construction Sector Council.

The NCC has 16 internal researchers and support personnel representing 30 FTEs working exclusively on research of the NIOSH Construction Research Program. Their research expertise includes economics, epidemiology, safety engineering, safety management, statistics, general engineering, industrial hygiene, science, medicine and toxicology. The CPWR has a consortium of academic and other institutions that have been awarded, through peer review, monies to work on a variety of construction-related projects.

Goals, Objectives, and Future Plans

The Construction Research Program focuses its activities on the achievement of four major goals:

- Goal 1: Reduce the major risks associated with traumatic injuries and fatalities in construction.
- Goal 2: Reduce exposures to health hazards associated with major risks of occupational illness in construction.
- Goal 3: Reduce the major risks associated with musculoskeletal disorders in construction.
- Goal 4: Increase understanding of construction sector attributes that affect occupational safety and health outcomes.

Two to six sub-goals are associated with each of the four major goals. The goals are a composite of goals and priorities that draw from NORA1, and internally generated strategic goals and high-priority topics.

The Construction Research Program proposes to continue with the current three-component organizational structure. There will be an open recompetition for the NCC in 2008-2009, and as in the past, scientific peer review will determine its award. The program expects that many of the current targeted areas requiring health and safety research will remain the same, but anticipates that some refocusing of program research efforts and emphases will be required by the influx of Hispanic workers into the construction industry, continuing changes in work organization, and the introduction of new technologies and materials such as nanoscale materials.

A draft of the second National Occupational Research Agenda (NORA2) was released for public comment in December 2007. The program intends to use the NORA2 Construction Sector Council goals for construction, combined with the findings and recommendations from this National Research Council review, to guide its next NCC Cooperative Agreement and to further strengthen the future program and its research directions.

EXTERNAL FACTORS

In accord with the Framework Document (Appendix A), the committee identified external factors that may have affected the Construction Research Program's relevance and impact during the 10-year period under review. The Framework Document addresses external factors as follows:

> External factors may be considered as forces beyond the control of NIOSH that may affect the evolution of a program. External factors influence NIOSH's progress through all phases of the logic model and flowchart, from inputs to end outcomes. . . . Identification of external factors by an EC [Evaluation Committee] is essential because it provides the context for evaluation of the NIOSH program. External factors may be best assessed on the basis of the expert judgment of EC members who have knowledge of the field of research. Information regarding external factors should also be sought from NIOSH, OSHA, and MSHA [Mine Safety and Health Administration] staff and from other stakeholders.

Several significant external factors affect both the relevance and the impact of the Construction Research Program, as described below.

Lack of Regulatory Authority

A fundamental external factor affecting the relevance and impact of the Construction Research Program is the fact that the program and NIOSH are research entities lacking regulatory authority. As such, the program can produce knowledge about construction workplace safety and health hazards and provide the application of this knowledge through a range of activities. Although the Construction Research Program can make recommendations to regulatory agencies, project owners, and contractors, how those recommendations are used, if at all, is beyond the control of the program.

Because NIOSH does not have standard-setting or enforcement authority, it relies on Congress and federal agencies such as OSHA, as well as on standards organizations, unions, and project owners and contractors both to implement and to enforce the outcomes and recommendations yielded by its research. For example, research conducted by the program and other organizations has identified effective interventions to protect construction workers from exposure to hazards such as crystalline silica and excessive noise. However, unless OSHA or another organization promulgates and enforces standards based on the available research, it is very difficult to link the impact of the program's research directly to reduced fatalities, injuries, and illnesses.

Methods other than standards for transferring research into practice are available, but they present similar issues of enforcement authority. For example, owners and clients who purchase construction services insert requirements in bid specifications to protect workers against unregulated as well as regulated hazards. Unfortunately, in the absence of such bid specifications and in the intensely competitive environment of construction contract bidding, those contractors who choose to devote resources to improving the health and safety of their workers by addressing unregulated hazards may be placed at a competitive disadvantage when bidding against contractors who do not choose to do so. Until the health and safety research from the program can be implemented through regulations promulgated by others or through changes in owners' and contractors' practices and behaviors, it will be very difficult to assess the degree to which the program's efforts have directly impacted worker health and safety on construction sites.

Resources

Total funding for NIOSH over the period FY 1996 through FY 2005 increased in absolute numbers, with a decrease in FY 2006. This increase is more modest once the absolute numbers are adjusted for inflation and for the increased costs of conducting scientific investigations due to new technologies and other factors using the Biomedical Research and Development Price Index (Table 2.1).

Over the same period, funding for the Construction Research Program has declined as a percentage of the overall NIOSH budget (Figure 2.4).

The committee believes that the level of resources available to the Construction Research Program has significantly limited the capacity of the program for conducting research activities across the broad range of safety and health hazards that are present on construction worksites. The lack of resources also limits the capacity of the program to conduct surveillance and training activities for the residential sector of construction with its preponderance of small firms located in

TABLE 2.1 Overall NIOSH Budget, FY 1996–FY 2006

	1996	1997	1998	1999	2000	2001	2002	2003	2004	2005	2006
Budget ($ millions)	161	173	184	204	226	260	276	273	277	286	255[a]
Adjusted (BRDPI)[b] (millions of 1996 $)	161	N.A.	N.A.	N.A.	199	221	227	217	212	211	182

NOTE: N.A., not available.

[a]In 2006, Congress redirected $35 million from the NIOSH budget appropriation to the Centers for Disease Control and Prevention for business support services.

[b]National Institutes of Health Biomedical Research and Development Price Index (BRDPI).

SOURCE: NIOSH (2007).

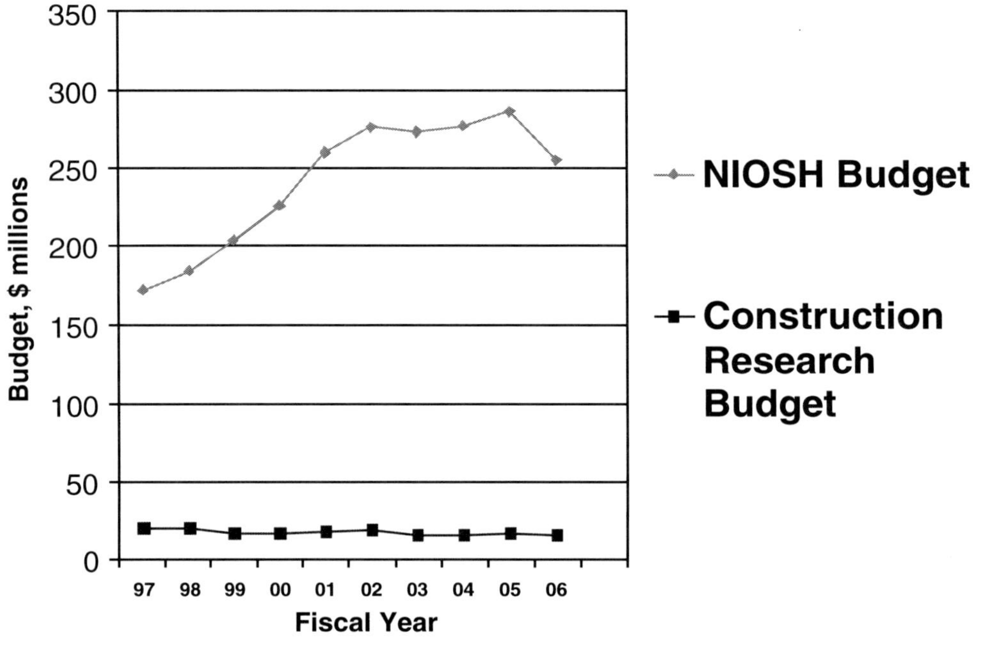

FIGURE 2.4 Comparison of NIOSH total budget and Construction Research Program budget in unadjusted dollars, FY 1996–FY 2006. SOURCE: NIOSH (2007).

every community in the country. The program needs to focus its R2P efforts on those sectors in which more formal mechanisms exist for reaching large contractors (heavy construction, industrial, commercial) and significant numbers of workers (labor unions).

A related issue is that until very recently, NIOSH senior management had not made the commitment to assign at least one full-time senior-level staff person to coordinate the array of projects and activities carried out within the Construction Research Program and one lead team member to serve as the Construction Program Manager. Over the review period, the Construction Research Program has primarily been under the direction of a steering committee of representatives from other program areas. There has been no single, senior-level person to advocate solely for the Construction Research Program and resources. The committee believes that the lack of full-time senior-level management has limited what could reasonably be accomplished by the program.

Industry Segmentation

As described in Chapter 1, the construction industry is highly segmented with respect to the number and wide variety of stakeholders involved, by the type of construction project, the physical environment in which it is built, how it is designed, the project's funding source (public versus private), and additional factors. This industry segmentation has significant implications for evaluating the impact of the Construction Research Program. First, it can be difficult to choose which research areas will have the greatest impact on the health and safety of the greatest number of workers, since safety and health performance across the different construction sectors varies significantly. This is primarily due to differences in the makeup of the workforce, exposures that workers may face on different types of projects, and the owner organizations. The segmentation also makes it difficult for individual researchers to translate research findings and promising health and safety interventions effectively and efficiently into practice across the industry as a whole.

Owners, especially those who are responsible for multiple projects or are in the public sector and are large contractors, can play an important role in construction health and safety by implementing safety practices and promoting a positive safety culture on their jobs and also by using contract language crafted to promote such an environment. In the union segment of construction, required union apprenticeship programs and established labor-management committees provide distinct structural vehicles for transferring worker training and engineering controls that generally do not exist in the non-union sector. Similarly, contractor associations and organizations offer an effective means for reaching non-union as well as union contractors. However, a large majority of construction firms are small operators with non-union, transient workforces. They may lack the resources to seek out the best practices or participate in training programs. Proactively disseminating research-based information to thousands of firms is challenging. Even when the information reaches such firms, there are few mechanisms for measuring the degree to which it has had any positive impact on health and safety.

The segmentation of the construction industry has also contributed to a related but distinct issue—that of engaging outside researchers interested in construction health and safety research in the program's extramural grants program. First, the complexity of the construction industry makes it difficult for researchers to access and then study specific and stable populations of construction workers or employers. Also, the inadequacy of surveillance data makes it somewhat difficult to ensure that research priorities reflect the most important issues. The Construction Research Program staff conveyed to the committee in its meetings, with some degree of frustration, that they have been unable to induce many extramural researchers to apply for grant monies to study how worker behavior and employer/management leadership skills can work together to promote a positive safety culture, a topic that is increasingly recognized as a key to best practices implementation.

REFERENCES

CPWR (Center to Protect Workers' Rights). 2002. Noise-induced hearing loss in construction. In: The Construction Chart Book, 3rd ed. Washington, D.C.: CPWR, Ch. 43.

CPWR. 2007. The Construction Chart Book: The U.S. Construction Industry and Its Workers. 4th ed. Washington, D.C.: Center for Construction Research and Training.

NIOSH (National Institute for Occupational Safety and Health). 2007. NIOSH Construction Research Program Evidence Package. Washington, D.C., July.

3

Evaluation of the Relevance and Impact of the NIOSH Construction Research Program

The committee was charged with reviewing and assessing the relevance and impact of the Construction Research Program of the National Institute for Occupational Safety and Health (NIOSH) in reducing construction-related workplace illnesses and injuries. The period selected for review was 1996 through 2005, to encompass 10 years of research following the program's first external review and the issuance of the first National Occupational Research Agenda (NORA1). As detailed in Chapter 2, the program's strategic planning and development of research goals have evolved over the period reviewed. In 2005, the Construction Research Program established a set of four goals and corresponding sub-goals (Box 3.1), which serve as the basis for the evidence package presented by NIOSH to the committee (NIOSH, 2007). This set of goals and sub-goals represents a composite of goals and priorities that were in place during the time frame of the review and which draw from earlier "High Priority Construction Topics" and the more recent draft NIOSH Strategic Goals.

To evaluate the relevance and impact of the NIOSH Construction Research Program, the committee divided into four teams of two or three members each, with each team assigned one of the four program research goals. Each team conducted an in-depth evaluation of the materials provided by the NIOSH Construction Research Program staff along with other information made available in subsequent meetings and communications with the staff. Following the guidance in the Framework Document (Appendix A), the committee carried out its evaluation using the terminology and organization of a logic model adopted by NIOSH to

BOX 3.1
Research Goals and Sub-goals of the NIOSH Construction Research Program

Goal 1: Reduce the major risks associated with traumatic injuries and fatalities in construction.

- 1.1 Falls from elevations
- 1.2 Contact with electricity
- 1.3 Struck-by incidents involving vehicles/equipment
- 1.4 Confined space, excavation, and trenching
- 1.5 Construction vehicle rollovers

Goal 2: Reduce exposures to health hazards associated with major risks of occupational illness in construction.

- 2.1 Reduce noise exposures and hearing loss
- 2.2 Reduce lead exposure and related health effects
- 2.3 Reduce silica exposure and silicosis
- 2.4 Reduce asphalt fume exposures and related health effects
- 2.5 Reduce dermal exposures and related skin disorders
- 2.6 Reduce welding fume exposures and related health effects

Goal 3: Reduce the major risks associated with musculoskeletal disorders in construction.

- 3.1 Reduce musculoskeletal disorders
- 3.2 Reduce disorders associated with excessive exposure to vibration

Goal 4: Increase understanding of construction sector attributes that affect occupational safety and health outcomes.

- 4.1 Use and improve surveillance resources to identify and track construction safety and health risks
- 4.2 Address special populations of employers and employees within construction (e.g., immigrant workers, youth workers)
- 4.3 Optimize the role of safety and health in construction training efforts
- 4.4 Explore promising approaches for addressing construction hazards
- 4.5 Improve diffusion of safety and health research to construction practice

SOURCE: NIOSH (2007).

characterize the steps in its work (Box 3.2). The evaluation process recommended in the Framework Document is illustrated in Figure 3.1.

The committee examined goals, inputs, activities, and outputs to evaluate the Construction Research Program's relevance in terms of its research priorities and the degree to which the program is engaged in appropriate transfer of activities for completed research projects and reported research results. Intermediate and end outcomes were the principal focus for the evaluation of the program's impact. The committee also considered the number of construction-relevant projects, the number of employees working on those projects, and the manner in which stakeholder input has been obtained and incorporated into program goal setting, strategic planning, and specific activities. The committee was particularly interested in the quantity and quality of program outputs (e.g., control technologies, guidelines, and education and training materials) and the degree to which those in the construction industry accepted and used Construction Research Program outputs.

ASSESSMENT OF RELEVANCE

In the following subsections, the committee highlights some, but certainly not all, of the activities undertaken by the Construction Research Program. The intent is to show the range of activities undertaken by both internal and external researchers. The presentation of the following material is organized by the goals and sub-goals as listed in Box 3.1.

Goal 1: Reduce the Major Risks Associated with Traumatic Injuries and Fatalities in Construction

Goals and Objectives

Within Goal 1, NIOSH identified five sub-goal areas for reducing fatalities and risks of injuries caused by safety-related (as opposed to health-related) hazards: Sub-goal 1.1, Falls from elevations; Sub-goal 1.2, Contact with electricity; Sub-goal 1.3, Struck-by incidents involving vehicles/equipment; Sub-goal 1.4, Working in confined space, excavation, and trenching; and Sub-goal 1.5, Construction vehicle rollovers. The committee received detailed information about each sub-goal, including project listings and other pertinent data from the NIOSH Construction Research Program staff. In addition, the committee used updated program information available online and relied on the experience and expertise of its members to assess the information and its pertinence to the committee's task.

NIOSH focused on this particular goal and sub-goals for several reasons. First, when compared with other industries, construction has a disproportionately high

BOX 3.2
Logic Model Terms and Examples

Planning Inputs: Stakeholder input, surveillance and intervention data, and risk assessments (e.g., input from Federal Advisory Committee Act panels or the National Occupational Research Agenda research partners, intramural surveillance information, Health Hazard Evaluations [HHEs]).

Production Inputs: Intramural and extramural funding, staffing, management structure, and physical facilities.

Activities: Efforts and work of the program, staff, grantees, and contractors (e.g., surveillance, health effects research, intervention research, health services research, information dissemination, training, and technical assistance).

Outputs: A direct product of a NIOSH research program that is logically related to the achievement of desirable and intended outcomes (e.g., publications in peer-reviewed journals, recommendations, reports, website content, workshops and presentations, databases, educational materials, scales and methods, new technologies, patents, and technical assistance).

Intermediate Outcomes: Related to the program's association with behaviors and changes at individual, group, and organizational levels in the workplace. An assessment of the worth of NIOSH research and its products by outside stakeholders (e.g., production of standards or regulations based in whole or in part on NIOSH research; attendance at training and education programs sponsored by other organizations; use of publications, technologies, methods, or recommendations by workers, industry, and occupational safety and health professionals in the field; and citations of NIOSH research by industry and academic scientists).

End Outcomes: Improvements in safety and health in the workplace. Defined by measures of health and safety and of impact on processes and programs (e.g., changes related to health, including decreases in injuries, illnesses, or deaths and decreases in exposures due to research in a specific program or subprogram).

External Factors: Actions or forces beyond NIOSH's control (e.g., by industry, labor, regulators, and other entities) with important bearing on the incorporation in the workplace of NIOSH's outputs to enhance health and safety.

SOURCE: Adapted from "Framework for the Review of Research Programs of the National Institute for Occupational Safety and Health," reproduced as Appendix A in this report.

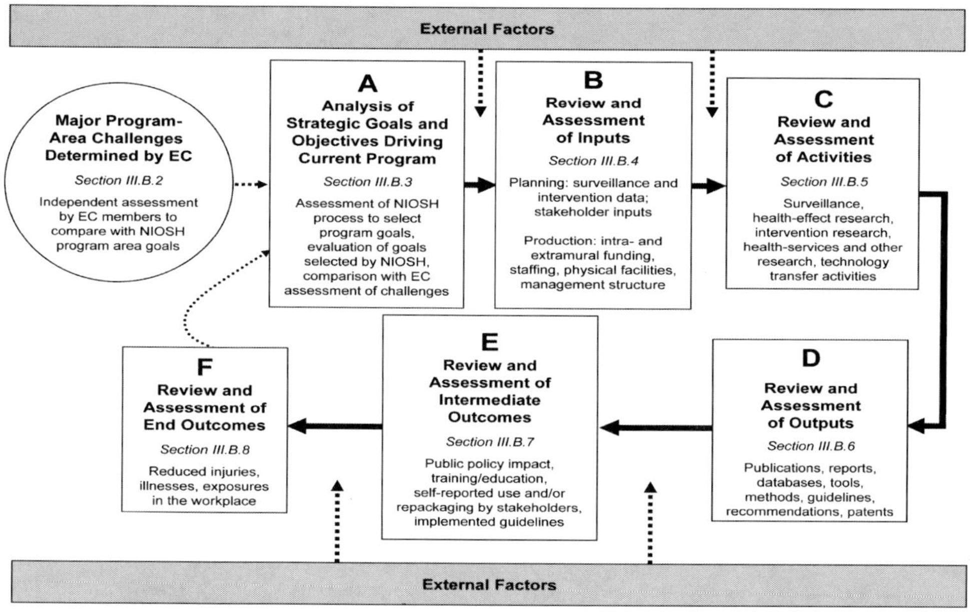

FIGURE 3.1 Flowchart of the evaluation process recommended by the Framework Committee. NOTE: EC, Evaluation Committee. SOURCE: Reprinted from "Framework for the Review of Research Programs of the National Institute for Occupational Safety and Health," reproduced as Appendix A in this report.

share of work-related fatalities. Second, the rate of fatalities in construction work has not declined as quickly as that of nonfatal injuries. Third, Congress directed NIOSH to focus on reducing construction fatalities. Finally, according to the presentation made and the supporting material within the evidence package (NIOSH, 2007), NIOSH theorized that by reducing fatalities, nonfatal injuries would also be reduced.

Planning and Production Inputs

The primary inputs to Goal 1 and its sub-goals are national and state surveillance data showing the causes of safety-related fatalities in the construction industry, which are (in order of frequency) falls from heights (35 percent), workers being struck by objects (12 percent), contact with electrical current (10 percent), workers struck by vehicle or equipment (6 percent), and workers caught in or crushed by collapsing materials (6 percent) (NIOSH, 2007). The Construction

Research Program seeks to reduce safety-related fatalities and injuries through four approaches:

1. *Hazard Identification*—Procedures or testing devices to determine that a hazard is present;
2. *Protection Equipment*—Equipment that prevents or minimizes injury if a worker encounters a hazard;
3. *Avoidance/Prevention*—Procedures or mechanical devices that cause the worker to avoid a hazard; and
4. *Behavior/Awareness*—Research aimed at preventing unsafe acts on a construction site.

Table 3.1 shows the distribution of research activities (by sub-goal) for each approach as identified by the committee. For example, the program's evidence package detailed eight research activities that were associated with Sub-goal 1.1 (falls from elevations). In examining these activities, the committee determined that three of the activities involved developing protection systems, four were focused on avoidance/prevention, and one was directed toward behavior/awareness. The committee used the same method for categorizing the research activities associated with the four other sub-goals: Sub-goal 1.2, Contact with electricity; Sub-goal 1.3, Workers struck by vehicles/equipment; Sub-goal 1.4, Confined space, excavation, and trenching; and Sub-goal 1.5, Construction vehicle rollovers.

TABLE 3.1 Program Distribution of Goal 1 Research Activities Focused on Reducing Safety-Related Hazards, by Sub-goals

	Sub-goal (Number of Research Activities)				
Focus of Research Activity	1.1	1.2	1.3	1.4	1.5
Hazard identification		2	1		
Protection equipment	3				1
Avoidance/prevention	4	1	2		2
Behavior/awareness	1	3	1	3	3

NOTE: Goal 1 and Sub-goals 1.1 through 1.5 are listed in Box 3.1 in this chapter.

Activities and Outputs

Sub-goal 1.1: Falls from Elevations

Falls from roofs, structural steel, scaffolds, ladders, and aerial lifts to lower elevations are the largest single cause of fatal injuries among construction workers. Falls account for one-third of construction fatalities, and construction fatalities from falls account for one-half of fall fatalities (394 of 770) across all industries (NIOSH, 2007). Falls are also a major cause of serious, nonfatal injuries.

Researchers in the Construction Research Program conducted a number of surveillance and investigation studies about fall injuries and fatalities, often learning where to focus prevention efforts in the process. They also conducted intervention research, training, and dissemination efforts. Over the period 1992 to 2005, program researchers authored a total of 29 peer-reviewed journal articles on this topic, provided 99 presentations, and developed 32 NIOSH and National Construction Center (NCC) publications, as well as 163 miscellaneous documents such as Fatality Assessment and Control Evaluation (FACE) reports, patents, and book chapters. Products related to program outputs have included adjustable guardrail assemblies, improved footware designs for work on roofs, and modified fall-protection harnesses.

NIOSH's research in the area of fatalities and injuries caused by falls from elevations has also included virtual reality technology. The Construction Research Program published validation studies for the surround-screen virtual reality (SSVR) system, the first SSVR system in the world designed for occupational fall-prevention research. The system is currently used to evaluate human performance at elevation, identify risk factors leading to fall incidents, and assess new fall-prevention strategies. Other work in this area has included publications addressing the effect of visual cues on balance control and research that addresses prototypes for sensory-enhancing technology to improve workers' balance (NIOSH, 2007).

Sub-goal 1.2: Contact with Electricity

Contact with electricity accounts for 10 percent of all construction-related fatalities. Construction accounted for almost half (47 percent) of all deaths associated with contact with electricity across all industries between 1992 and 2002 (Cawley and Homce, 2006). Deaths and injuries due to contact with electricity occur not just among electricians, but also among roofers, painters, laborers, operating engineers, and carpenters, all of whom may work near overhead power lines (OHPLs), wiring, transformers, light fixtures, machines, and power tools.

In the area of contact with electricity and electrocutions, NCC researchers conducted a survey of work practices among 5,000 International Brotherhood of Electrical Workers electricians. Failure to lock out or tag out electrical equipment before beginning to work is a major cause of construction electrocutions. Researchers in the Construction Research Program have also been evaluating the performance of OHPL proximity-warning alarm devices on mobile cranes. The test protocol for the evaluation was developed in conjunction with a number of partners including the International Union of Operating Engineers (IUOE), the Center to Protect Workers' Rights (CPWR), Occupational Safety and Health Administration (OSHA), Zachry Construction Corporation, Allied Safety Systems, Inc. (manufacturer of Sigalarm), Allied 99 Safety Engineering, Inc. (OHPL alarm manufacturer), and the Association of Equipment Manufacturers. Full-scale electrical tests were executed using a mobile crane outfitted with the OHPL alarm devices from two manufacturers. The tests quantified the distance from a power line at which each alarm device sounded a warning in various crane-and-power line configurations (NIOSH, 2007). Construction Research Program staff were also awarded a patent for an OHPL-contact alarm system.

Sub-goal 1.3: Struck-by Incidents Involving Vehicles/Equipment

Struck-by incidents can involve workers struck by vehicles, mobile equipment, or falling and flying objects. Almost half of the 802 construction workers killed by vehicles or equipment between 1995 and 2002 worked on highway and street construction projects (BLS, 2007). Falling objects can include wrenches or other equipment and tools, whereas flying objects include nail guns and other power tools. More than 22,000 workers per year are treated in hospital emergency rooms for nail gun injuries (CDC, 2007).

In the area of workers struck by vehicles/equipment, program researchers have conducted 53 highway and street construction Fatality Assessment and Control Evaluation investigations. In 2000, they began evaluating interventions involving a variety of proximity-warning systems that provide construction equipment operators the ability to monitor blind spots. They have also evaluated administrative approaches for controlling construction vehicle and worker movements within work zones (referred to as internal traffic control plans, or ITCPs). The program has disseminated a significant amount of research information to owners' and workers' organizations and associations, including the Washington State Department of Transportation, American Road and Transportation Builders Association, IUOE, National Asphalt Pavement Association, and OSHA. The disseminated information has included information from research for the development of standards and relevant materials incorporated into course curricula for construction engineering and management. NCC research has included surveillance efforts to contribute to the understanding of factors related to safety during nighttime construction in work zones. Results from these surveillance activities are used to identify new risk factors, identify injury-prevention strategies, and guide and prioritize future research efforts.

Sub-goal 1.4: Confined Space, Excavation, and Trenching

Between 1992 and 2001, approximately 540 construction workers died in trenching or excavation cave-ins (MMWR, 2004). Between 1997 and 2001, an additional 89 workers died in confined spaces, such as storage tanks, pits, boilers, ventilation and exhaust ducts, sewers, tunnels, pipelines, and underground utility vaults (Meyer, 2003).

NIOSH has a long-standing interest in safety risks associated with construction work in trenches, excavations, and confined spaces. NIOSH issued a recommended standard for working in confined spaces in 1979, issued safety guidance in 1987, and provided technical input to OSHA for the revision of its excavation standard in 1989. The program also provided input in 1995 to OSHA's general industry standard for confined spaces.

Program researchers have used surveillance data from a variety of sources to identify activities, trades, and risk factors regarding fatal and nonfatal injuries in trenches and confined spaces. They provided support for ongoing efforts to develop pocket guides to chemical hazards commonly used in confined spaces. They have prepared 30 publications, and products, and provided training for 6,700 workers in 31 states. The Construction Research Program has undertaken a number of safety interventions for excavation and trenching, including teleoperation of mechanical devices to dig trenches and install pipe, the trench box safety project, and safety in trench operations.

Program researchers also assisted in the design of a compact disc (CD)-based training module to raise awareness of trenching hazards. Later they modified the CD on the basis of feedback from decision makers, trainers and consultants, and workers in the construction industry.

Sub-goal 1.5: Construction Vehicle Rollovers

Heavy equipment—cranes, excavators, tractors, loaders, bulldozers, pavers—are commonly used on construction projects. Vehicle overturns or rollovers are among the most common causes of construction fatalities associated with vehicles and equipment.

For vehicle rollovers, NIOSH conducted a number of investigations and used existing surveillance resources and programs to identify risk factors associated with construction vehicle overturns and to develop prevention measures to address them. This work yielded, among many things, guidance titled "NIOSH Alert—Preventing worker injuries and deaths from mobile crane tip-over, boom collapse, and uncontrolled hoisted loads" (as indicated in NIOSH, 2007). The dissemination of this alert was combined with that of numerous publications and presentations.

Goal 2: Reduce Exposures to Health Hazards Associated with Major Risks of Occupational Illness in Construction

Goals and Objectives

Goal 2 pertains to eliminating or at least mitigating health hazards on construction sites. Its six specific sub-goals target the following: Sub-goal 2.1, Reduce noise exposures and hearing loss; Sub-goal 2.2, Reduce lead exposure and related health effects; Sub-goal 3.3, Reduce silica exposure and silicosis; Sub-goal 2.4, Reduce asphalt fume exposures and related health effects; Sub-goal 2.5, Reduce dermal exposures and skin disorders; and Sub-goal 2.6, Reduce welding fume exposures and related health effects.

The Centers for Disease Control and Prevention (CDC) has defined two types of exposure: *acute exposure* is exposure to a specific chemical for 14 days or less; *chronic exposure* is exposure to a chemical for 365 days or more (CDC, 2008). For construction research, the program has further defined these terms to note that chronic exposures happen not just over time but also across multiple worksites, causing a cumulative effect with potentially negative physical reactions revealing themselves after a particular project is completed. In this situation, it is not always one particular chemical or substance that is the cause, but rather it is the cumulative effect of the exposure over time. In the case of acute exposures, any resulting negative effects occur more quickly and are more easily connected to a particular chemical or substance.

Planning and Production Inputs

Unlike the available statistics for safety-related hazards, meaningful national statistics are not available to identify the leading causes of workplace-related diseases or deaths and to help target health-related research and interventions. It is in fact well established that work-related illnesses are very difficult to recognize and often remain unreported or underreported on OSHA logs and thus are not captured by the Bureau of Labor Statistics (BLS) annual survey. This underreporting problem masks the true occurrence of occupational illness in construction.

Researchers in the Construction Research Program have relied on a variety of sources, including national and state-level illness statistics, knowledge and extrapolation of construction exposures, studies based on information on death certificates, and international surveillance to identify and characterize many of the health hazards present on construction sites. Once such hazards are identified, program researchers have engaged in detailed planning efforts with construction industry stakeholders to prioritize the needs as well as to help the program develop, implement, and evaluate interventions to eliminate or at least mitigate health outcomes from exposures that are health hazards. Stakeholder input has been gathered by program researchers using multiple methods, including sponsoring and/or attending construction safety and health national conferences and initiating and developing partnerships with contractors, contractor organizations, labor unions, and equipment manufacturers. For example, lead exposure and its adverse health consequences together with noise exposure and hearing loss were identified as important issues by stakeholders at national conferences sponsored by the Construction Research Program. Program surveillance data supported the identification of these issues as a priority for research.

Activities and Outputs

Sub-goal 2.1: Reduce Noise Exposures and Hearing Loss

NIOSH has a distinct research program for hearing loss across all industries that has its own set of goals and sub-goals, funding, and staffing. The Institute of Medicine's review of the Hearing Loss Research Program noted that "research efforts [addressing engineering controls to reduce noise exposure] have been concentrated in the mining sector, with some attention to the construction sector" (IOM, 2006, p. 6). One of the Hearing Loss Research Program goals is to develop engineering controls to reduce noise exposure. Two of the sub-goals are to reduce noise generated by roof bolting machines using wet and mist drilling and to reduce noise exposures to construction workers using a Web-based database for powered hand tools. Although the Hearing Loss Research Program has developed the database, it has not translated the research into engineering noise controls (IOM, 2006, p. 10). The evaluation committee for the Hearing Loss Research Program also stated that "noise control engineering should be the primary approach to prevention of hearing loss" (IOM, 2006, p. 17).

Within the construction industry, all workers are at risk for exposure to harmful levels of noise from heavy equipment and power tools. Noise exposures in construction tend to be highly episodic, with relatively short duration but extremely high peak characteristics (referred to as *impulse noise*), for example, exposure to jackhammers. Workers can be exposed to such noise while acting as operators of vehicles and equipment or simply while being in proximity to the use of vehicles and equipment. One study estimates that construction workers in all trades have a 60 percent probability of developing hearing loss (Dement et al., 2005).

In the area of noise exposure and hearing loss, the Construction Research Program has developed and implemented engineering controls and work practice interventions, as well as education, training, and guideline materials to increase workers' use of hearing protection and contractor implementation of hearing-protection programs. Engineering noise control efforts have been focused on powered hand tools and to a lesser extent on drill rigs. Program researchers have published numerous peer-reviewed articles and made presentations at conferences addressing the prevalence of noise exposure on construction sites and its potential to result ultimately in hearing loss. In addition, they have worked closely with OSHA's Advisory Committee on Construction Safety and Health to advise OSHA to move forward with promulgating a revised hearing-conservation standard for the construction industry. OSHA has issued an advanced notice of proposed rule making to revise the existing standard, but the standard has not yet been revised. The program uses the NIOSH and NCC Web sites, in particular, eLCOSH,[1] to disseminate construction-related health and safety materials.

Sub-goal 2.2: Reduce Lead Exposure and Related Health Effects

Construction workers, particularly those involved with remodeling, repair, demolition, or remediation projects, continue to be exposed to lead. Plumbers, painters, electricians, and welders are among the trades exposed to lead and at risk for lead poisoning. NIOSH has been involved with activities to reduce exposures to lead for all industry workers since the 1970s. The Construction Research Program contributed research used in the development of a lead-in-construction standard issued by OSHA in 1992 (NIOSH, 2007).

During the review period, the Construction Research Program has developed research and analytical methods for characterizing lead exposures. Program researchers have collaborated with the Adult Blood Lead Epidemiology and Surveillance (ABLES) program and provided technical advice to the U.S. Environmental Protection Agency (EPA) and the U.S. Department of Housing and Urban Development (HUD) on guideline documents and have proposed rules that address lead-assessment and lead-abatement projects.

The program has also made a major contribution to the issue of take-home exposures of lead from construction activities. Program-generated research has

[1]The electronic Library of Construction Occupational Safety and Health—eLCOSH—is an electronic clearinghouse for construction-specific safety and health information. Available at http://www.cdc.gov/elcosh/. Accessed August 15, 2008.

been used by a number of states for their public health programs to reduce lead exposures to construction workers and their families.

The program also helped to develop and evaluate portable methods for identifying lead at the workplace, in the air, in surface dust, and in workers' blood levels. Program researchers developed and patented a colorimetric wipe that quickly and easily detects the presence of lead on skin and surfaces. The technology was licensed and is sold commercially.

Sub-goal 2.3: Reduce Silica Exposure and Silicosis

Silicosis is a disabling and sometimes fatal lung disease caused by breathing dust that has very small pieces of crystalline silica in it. Crystalline silica is found in concrete, masonry, sandstone, rock, paint, and other abrasives. The cutting, breaking, crushing, drilling, grinding, or abrasive blasting of these materials may produce fine silica dust. It can also be in soil, mortar, plaster, and shingles. A project to help understand the exposures of construction workers to silica was one of the original projects undertaken by the Construction Research Program.

To reduce exposure to silica and prevent silicosis, program researchers have established collaborative relationships with other government agencies through an interagency working group addressing silicosis prevention. This working group includes representatives from the Mining Safety and Health Administration, OSHA, the U.S. Geological Survey, and the Bureau of Mines (now part of NIOSH). Researchers have developed an analytical method for measuring silica, published 41 peer-reviewed journal articles on silica exposure, and made numerous presentations of their findings at conferences and meetings. In 2002, the program published a silica-hazard policy review that has been circulated widely and is available on the NIOSH Web site (http://www.cdc.gov/niosh/). The document should be quite useful to OSHA if it moves forward on its regulatory agenda to develop a proposed standard for respirable crystalline silica in construction.

Sub-goal 2.4: Reduce Asphalt Fume Exposures and Related Health Effects

In the construction industry, asphalt is primarily used in roofing and road-building projects. The heating of asphalt during application results in the release of more than 50 organic compounds to which 350,000 construction workers are routinely exposed. As with lead exposures, NIOSH has conducted research on asphalt exposures since the 1970s. In testimony to OSHA in 1988, NIOSH recommended that asphalt fumes be considered as a possible occupational carcinogen (NIOSH, 2007).

During the review period, program activities on asphalt fume exposure have been quite extensive. A substantial and effective partnership was developed—program partners on this initiative included the Federal Highway Administration, the National Asphalt Pavement Association, the Laborers' International Union of North America, the IUOE, and the Asphalt Institute—that focused on prevention research to reduce worker exposure on asphalt paving projects throughout the United States. Working with highway-class paver manufacturers representing more than 80 percent of the asphalt paver manufacturing market, researchers assisted them and the stakeholders in a coordinated and planned effort to design and field-test engineering controls for paver equipment that reduced worker exposure to asphalt fumes. Protocols were developed to assess the effectiveness of engineering controls, and performance evaluations of those controls were tested in the field. In a voluntary agreement between industry, organized labor, and OSHA, paver manufacturers committed to incorporating engineering controls on all pavers manufactured after 1997.

For roofing applications of hot asphalt, program researchers developed a work practice guidance document designed to reduce the exposure of roofers to asphalt fumes. The work practice advice was incorporated into training programs by both labor and management organizations.

Work on methods research was designed to develop laboratory techniques to generate asphalt fumes that are similar to fumes generated by construction applications in the field. Results from this work were then linked with a research protocol prepared by the National Toxicology Program for assessing subchronic noncarcinogenic responses to asphalt in rats and potentially for use in a long-term carcinogenicity study.

In response to health concerns regarding the federal legislative requirement to use crumb rubber modifiers in highway paving projects, NIOSH researchers characterized and compared field exposures and health effects of conventional asphalt and asphalt modified with the addition of crumb rubber.[2]

The program's research on health effects of asphalt has resulted in relevant outputs, including the work on controlling exposures to roofing asphalt fumes. Program researchers published a hazard review document on asphalt that, in addition to summarizing health data from animal and human studies, outlined additional research needed to reduce worker exposures. An acute-irritant health-effects study of road-paving workers, conducted by program researchers, was also published.

Sub-goal 2.5: Reduce Dermal Exposures and Related Skin Disorders

A number of chemicals used in the construction industry can cause skin disorders, such as contact dermatitis, a painful and sometimes debilitating disease.

---

[2]Crumb rubber is recycled rubber from automotive and truck scrap tires from which steel and fluff are removed during the recycling process. A crumb rubber modifier is a machine that converts this crumb rubber into a consistency that can be sprayed.

The chemicals include epoxies, solvents, preservatives, and Portland cement. For Portland cement alone, it is estimated that more than 1.3 million construction workers are employed in occupations with exposures to wet cement, making it one of the most frequently encountered dermal hazardous substances (CPWR, 1999). Examples of trades with potential exposure include bricklayers, cement masons, concrete finishers, construction craft laborers, hod carriers, plasterers, terrazzo workers, and tile setters.

Construction workers are also at significant risk for developing skin cancer due to prolonged exposure to sunlight (Espinosa et al., 1999).

There have been numerous research activities and outputs related to dermal exposure and reducing skin disorders. The program identified best practices and developed a number of documents particularly focused on Portland cement. Program researchers assisted NIOSH in developing its testimony for the OSHA proposed rule on hexavalent chromium, a component in Portland cement that can cause allergic dermatitis. The program, working with its partner organizations, produced a useful and relevant series of manuals, pamphlets, hazard alerts, and a training PowerPoint® presentation that addresses protecting workers while working with wet cement. These documents have been widely distributed and are also available on the eLCOSH Web site for downloading.

Sub-goal 2.6: Reduce Welding Fume Exposures and Related Health Effects

It is estimated that more than 410,000 workers weld, braze, cut, or solder full time, with more than 1 million workers welding on an intermittent basis (NIOSH, 2007). The International Agency for Research on Cancer has concluded that welding fumes were "possibly carcinogenic" to humans. Several epidemiological studies suggest that exposure to welding fumes is associated with an increased incidence of lung cancer (Society of Toxicologic Pathology, 2006) and other respiratory illnesses, including asthma, bronchitis, and chemical pneumonitis.

With respect to reducing or eliminating welding fume exposures, program researchers have carried out numerous activities including the development of new methods for identifying hexavalent chromium (one of the many hazardous components of welding fumes) in workplace air and for assessing the effectiveness of control measures in reducing exposures in welders. The program has published 45 peer-reviewed articles and contributed numerous national and international conference presentations addressing various topics pertaining to welding fume reduction. NCC researchers have used task-based exposure methods to characterize welding fume exposures and have identified and evaluated engineering controls (e.g., mechanical ventilation).

Goal 3: Reduce the Major Risks Associated with Musculoskeletal Disorders in Construction

Goals and Objectives

NIOSH has two sub-goals for research Goal 3: Sub-goal 3.1, Reduce musculoskeletal disorders (MSDs);[3] and Sub-goal 3.2, Reduce disorders associated with excessive exposure to vibration. The objective of the first sub-goal is to reduce the incidence and severity of MSDs that arise from some combination of physical force, performing tasks involving repetitive motions, performing tasks in awkward or static postures, performing tasks with extensive vibration, or working in harsh environments. The second sub-goal pertains to reducing damage to soft tissues caused by work with powered hand tools, such as chain saws, drills, and riveters, and by construction vehicles.

Planning and Production Inputs

A major planning input for this goal area has been the surveillance data gathered from a variety of sources, as well as surveillance data that the Construction Research Program collects as part of its overall activities. A significant part of program efforts, particularly in the early stages, was to gather better data to help define the scope and severity of MSDs in construction workers. Program researchers and the NCC have also involved contractors and workers at the initiation stage of research to evaluate proposed interventions to ensure that they are appropriate for construction trades and tasks.

Although it is well known that MSDs are common, it has been difficult to develop national estimates of work-related MSDs. Program research has included cross-sectional surveys of construction workers in several states, active surveillance systems based in emergency rooms to improve the assessment of the incidence of injuries, and the use of BLS databases as well as other national databases. This research demonstrated that work-related MSDs, including carpal tunnel syndrome, were more common than indicated by national data, that the pattern of disorders varies among trades, and that such disorders are frequently chronic or recurrent and have significant impact on construction workers' quality of life.

Inputs to vibration-related disorders have included the report *Musculoskeletal Disorders and the Workplace: Low Back and Upper Extremities* (NRC, 2001), which described gaps in current research that should be addressed. Based on this report, the program addressed testing systems for vibration exposure from hand tools and equipment and assessed the health effects of vibration exposures. Stakeholder

[3]Injuries or illnesses of the muscles, tendons, joints, and nerves caused or aggravated by work.

input relevant to MSD research has also occurred through program-sponsored conferences that explicitly incorporated ergonomics as well as other safety and health topics.

Exposure assessment has used observational, qualitative, and quantitative studies to characterize exposures in a variety of trades, including masonry, ironwork, drywall installation, carpentry, and sheet metal work. In addition to collecting exposure data using existing methods, researchers have developed new exposure-assessment and data-analysis methods. Work on exposure to vibration is driven by data showing that more than half a million construction workers are exposed to whole-body vibration and that the highest percentage of excessive hand-transmitted vibration exposure occurs in the construction sector. MSD research has also explicitly involved economic (cost) analysis to assess potential solutions, their diffusion into the workplace, and their impact on worker productivity.

Activities and Outputs

Activities and outputs related to Goal 3 and its sub-goals have included five new exposure-assessment methods and two new data-analysis methods relevant to calculating exposure to physical risk factors for MSDs. Approaches taken by program researchers to address MSDs included observational, qualitative, and quantitative studies to characterize exposures in highway construction, masonry, ironwork, and drywall installation.

Intervention activities have included participatory ergonomics and training interventions in multiple trades, including concrete construction, carpentry, scaffold erection, and on large, multiyear construction projects. Activities have also included more than 25 evaluations of specific tools and equipment for working overhead, working at foot level, and performing hand-intensive work. A large number of educational materials aimed at workers and contractors have been developed or funded by the program and are made available at no cost in print and online. These include the "Bright Idea" series, the "Sensible Solutions" series, and documents that describe successful interventions designed to prevent MSDs in multiple trades, including drywall installation, the electrical trades, the mechanical trades, and ironwork.

Conference sponsorship and stakeholder meetings have been an important program output, with multiple conferences designed to provide content on MSD prevention and targeted to contractors and unions. For example, one conference organized by program researchers in conjunction with the International Ergonomics Association led to the development of research tools and the sharing of findings and experiences. Bilateral cooperation with European countries including Sweden and the Netherlands has resulted in ergonomic tool design, technology design, and approaches to industrywide solutions.

Approaches taken by program researchers to meet Sub-goal 3.2 include determining methods for reducing vibration at the source, reducing vibration transmitted to the human body, monitoring and controlling the duration of vibration exposure, developing new methods for quantifying vibration exposure on the basis of location-specific biodynamic responses, developing effective methods to take into account the effects of hand forces and postures in dose quantification, examining the responses and health effects of vibration, and examining hands-on vibration syndrome through the development of new or the improvement of existing testing methods and procedures.

Program outputs related to vibration-related disorders have included more than 50 journal articles, more than 60 conference presentations, and three awards for research, including a 2006 Liberty Mutual Award and the 2005 and 2007 NIOSH Alice Hamilton Awards for best published articles. Technical transfer activities include the initiation of a new series of conferences called the American Conference on Human Vibration. The first such conference was conducted by the program in Morgantown, West Virginia, in the summer of 2006. It brought together researchers from many countries and from across the United States to discuss issues related to whole-body and hand-transmitted vibration. The program has sponsored more than five invited technical seminars. It has provided consultations to public stakeholders related to vibration exposure and has provided health-hazard evaluations for employers.

Goal 4: Increase Understanding of Construction Sector Attributes That Affect Occupational Safety and Health Outcomes

Goals and Objectives

There are five sub-goals associated with research Goal 4: Sub-goal 4.1, Use and improve surveillance resources to identify and track construction safety and health risks; Sub-goal 4.2, Address special populations of employers and employees within construction (e.g., immigrant workers, youth workers); Sub-goal 4.3, Optimize the role of safety and health in construction training efforts; Sub-goal 4.4, Explore promising approaches for addressing construction hazards; and Sub-goal 4.5, Improve diffusion of safety and health research to construction practice.

The overall nature of Goal 4 and its sub-goals is different from that of the other three goals in that Goal 4 and its associated sub-goals are not research goals per se. Rather they represent overarching programmatic issues, of which Sub-goals 4.3 and 4.4 were addressed in the review and discussion of Goals 1, 2, and 3. Therefore, the committee focused its relevance and impact assessments on Sub-goals 4.1, 4.2, and 4.5.

Planning and Production Inputs

Surveillance methods have been and continue to be key to the successful conduct of the program's research related to special populations. Demographics over the past 10 years have changed significantly, and the increase of Hispanic workers within the construction industry has been fourfold (NIOSH, 2007). Combined with this change is the increase of Hispanic workers in non-union positions and residential construction positions, making it a greater challenge to conduct effective surveillance.

During the committee's interviews with stakeholder panels (see Appendix B), a particular challenge that emerged was the accuracy of data pertaining to special populations. Data accuracy for young workers and women was believed to be less problematic than that for immigrant and Hispanic segments of the worker population. Clearly the industry has experienced a large influx of new workers, many of whom rotate from project to project on a daily basis. This rotation has created an environment in which it has been difficult to obtain stable and hence accurate data. In addition, many of these workers have functioned as independent contractors, operating outside any system that would allow a more accurate recording of surveillance data. In the committee's examination of how the Construction Research Program has conducted its surveillance of special populations, the relationship developed by NIOSH researchers with various state and federal agencies and academic institutions to develop a national construction surveillance capacity was evident.

Activities and Outputs

Sub-goal 4.1: Use and Improve Surveillance Resources to Identify and Track Construction Safety and Health Risks

NIOSH staff and the NCC have collaborated to use and improve surveillance research resources. The purpose of health and safety surveillance research in construction is to give the program a sense of direction and a notion of priorities for study purposes. The foundational activity in construction safety surveillance research has been to consolidate, as much as feasible, all available construction-focused safety and health data from national, state, local, public, and private sources. This compilation of research has allowed researchers quicker access to relevant data, provided the public with health- and safety-related data on the industry, and helped to identify gaps in the available research. Thus, identifying available research has not only assisted in guiding construction-focused health and safety research but also has spurred other research entities to develop or alter their data-collecting activities in ways that will improve construction safety surveillance.

Sub-goal 4.2: Address Special Populations of Employers and Employees Within Construction

The construction workforce includes three groups with unique needs or disproportionate risks: Hispanics, women, and young workers. Of these, the group comprising Hispanic workers (29.3 percent of the construction workforce) is several orders of magnitude larger than that of women (9.6 percent) or of workers ages 15 to 17 years (3.7 percent). Two-thirds of the women are involved with construction management or technical and support positions, while only one-third are in production-related activities. Young workers are of concern because they represent 18 percent of construction-related fatalities but less than 4 percent of all construction workers.

The Construction Research Program has published 25 peer-reviewed journal articles, made 13 presentations, and produced 19 NIOSH and CPWR publications to address the goals and objectives for Sub-goal 4.2. The principal thrust of these activities has included educating policy makers and regulators (briefings to Congress, advisories to the Department of Labor Employment Standards Committee, presentations and publications on health and safety concerns of construction tradeswomen); collaborating with organizations that have influence within and disseminate information to the construction industry (participation in the OSHA Hispanic Task Force, participation in National Research Council workshop Safety Is Seguridad), and establishing a direct outreach to production workers within these special populations (eLCOSH Spanish language materials, interviews on National Public Radio, a national conference on immigrant workers, distribution of 8,000 copies of pamphlets containing information on young workers' rights to small contractors nationwide). Of these activities, the majority have focused on Hispanic workers.

With regard to Hispanic workers, specific objectives identified by the program included (1) examining the safety and health status of Hispanic construction workers, (2) identifying disparities in safety and health and the utilization of health services among Hispanic construction workers, (3) identifying major socioeconomic and work organization factors contributing to the disparities and creating potential barriers to injury-reducing interventions, and (4) developing intervention strategies to reduce or eliminate the disparities and improve the safety and health of Hispanic construction workers overall.

The committee noted that the evidence package provided to the committee by the NIOSH program staff referred to both "Hispanic" and "immigrant" workers (NIOSH, 2007). However, through most of the evidence package, most references were made to "Hispanic" workers, with few references to "immigrant workers."

NIOSH states the following in its online documentation (http://www.cdc.gov/elcosh/docs/other/women_minorities.html):

> Hispanic construction workers are approximately one third native-born and two-thirds foreign-born workers. While unauthorized immigrant workers represented 4.9% of the total civilian labor force, they represented 14% of the construction workforce. Hispanics constituted 66.5% of growth in the construction workforce in 2005-6.

It is unclear from the material in the evidence package just how NIOSH is approaching its surveillance of special populations. Interventions targeted at workers of Hispanic descent clearly identify one particular ethnic population, a sizable part of the construction industry. In other parts of the evidence package, the program refers to "immigrant workers," a term with broader connotations. It appeared that "Hispanic" and "immigrant" were used synonymously. The introduction of "unauthorized immigrant workers" raises a much larger and more problematic set of issues that in fact go beyond NIOSH's ability or authority to address.

Construction Research Program and NCC researchers have focused on and conducted projects to examine the health and safety of Hispanic construction workers and identify disparities in safety and health and utilization of health services by these workers. In addition, researchers have focused on identifying potential barriers to injury-reducing interventions and as a result have developed intervention strategies to reduce and/or eliminate the disparities and improve the safety and health of Hispanic construction workers overall. This research has included investigations of safety culture, work practices, risk perception, and length of time in the United States and the development of culturally tailored OSHA training modules and evaluation of their impact. This remains a challenging area for the program given the dynamic nature of this specific subpopulation's work profile.

Sub-goal 4.5: Improve Diffusion of Safety and Health Research to Construction Practice

The terms "diffusion," "dissemination," and "technology transfer" all refer to how information, products, and ideas are spread from the sources where they were created to others who might benefit from having them or at least knowing about their existence. At NIOSH, and more specifically within the Construction Research Program, these activities as a whole have been referred to as research-to-practice, or R2P, and they cut across all of the program's goals and sub-goals. R2P is used here as an umbrella term encompassing these types of activities. Box 3.3 contains the goals of the overall NIOSH R2P program along with illustrative actions taken by this program.

BOX 3.3
NIOSH Research-to-Practice (R2P) Goals and Actions

R2P Goals

- Develop effective products,
- Translate research findings into practice,
- Target dissemination efforts,
- Evaluate and demonstrate effectiveness of these efforts in improving worker health and safety.

R2P Actions

Creation of Guidance for Reducing Hazards in the Tower-Erection Industry

The National Institute for Occupational Safety and Health (NIOSH) worked with the Occupational Safety and Health Administration and the National Association of Tower Erectors on a multiagency task force to address hazards associated with telecommunications tower construction and maintenance.

Reducing Worker Exposure to Asphalt Fumes

A unique government, industry, and labor partnership was formed to reduce worker exposure to asphalt fumes during paving operations. This partnership was successful in developing practical, effective control systems to control asphalt fume exposures.

Creating a "Construction Solutions" Database

NIOSH partner the Center to Protect Workers' Rights is currently developing an online "Construction Solutions" database to organize hazards by tasks and allow workers and contractors access to options for controlling those hazards.

SOURCE: NIOSH (2007).

Activities and Outputs

The R2P activities carried out to date by program researchers vary greatly, depending on what was diffused or disseminated (i.e., outputs), to whom it was provided (i.e., target audience), and how it was given out (i.e., methods and processes used). The majority of R2P program outputs were educational materials in either print or electronic format. They include scientific articles published in peer-reviewed journals or information on controls and/or solutions to specific construction hazards presented as videos, CDs, brochures, hazard alerts, and criterion documents. Engineering controls were also diffused (some in partnership with industry and labor) such as asphalt-paver retrofits to reduce and/or eliminate fumes, and leading-edge fall-protection systems. Program researchers have used national conferences and stakeholder consultations to help determine the best format and processes for information dissemination. The target audiences for R2P activities are just as varied as the outputs. They include but are not limited to workers, employers, other researchers, safety professionals, and government agencies.

R2P efforts are either passive or active. Passive methods entail having members of the target audience come to the source (or its proxy) to gain access to and/or retrieve the desired output. Active methods "push" the information out to audiences. Both types are needed for an effective overall R2P strategy.

In terms of passive methods, the Construction Research Program has used primarily three sources: (1) the NIOSH Publications Office, (2) eLCOSH, and (3) the Construction Solutions database. The NIOSH Publications Office has been available to the public for many years. The public can call, write, e-mail, or send a message via the NIOSH homepage (www.cdc.gov/NIOSH) to the office in Cincinnati, Ohio, requesting one or more copies of any publication developed by NIOSH researchers. Prior to 2000, the program did not collect standardized data to document the publication name, quantity requested, and requestor's name. Since 2000, implementation of standard processes has resulted in more useful data, which paint an impressive R2P picture. Looking specifically at 30 construction-related documents covering a broad range of topics (e.g., electrocutions, falling from heights, carbon monoxide poisoning, asphalt fume exposure, and drywall dust exposure) prepared between 1985 and early 2007 for which dissemination data are available, well over 250,000 documents have been requested by and distributed to a broad spectrum of individuals and entities both nationally and internationally.

Examples of domestic requests for these documents made by both union and non-union representatives are those from federal agencies (Army Corps of Engineers, U.S. Department of Labor/OSHA), from state organizations (State of Nevada–Business and Industry, Michigan OSHA, Minnesota Department of Health, New York State Department of Health), and from local governmental agencies (Miami Dade Building Code Compliance Office; City of Wheeling, West Virginia). Requests have come from large companies (Shell Oil Company, ConAgra Foods, Cargill, Huber Hunt and Nichols Builders, the Lockheed Martin Corporation) and small companies (Terry's Electric, Brown Printing Company, Nickles Bakery, Nobis Engineering, Feather Falls Casino); from colleges and universities (Texas A&M University, Fullerton College, Keene State College), from public schools (Mankato

Area Schools), and from vocational schools (Quincy Public School–Technical Education, James D. Patten Vocational School); and from employers (Association of General Contractors, Construction Employers Association, Builders Exchange of Central Ohio), labor (International Brotherhood of Electrical Workers Local 212 and Local 725, Ohio Operating Engineers, Laborers Health and Safety Fund of North America, Chicago Women in the Trades), and trade organizations (National Asphalt Trade Association). Other requests have come from insurance companies (The Hartford, Cincinnati Insurance) and consulting companies (Jaggers Safety Consultants, Safety Priority Consultants).

The electronic Library of Construction Occupational Safety and Health (eLCOSH) is an electronic clearinghouse for construction-specific safety and health information. This site records more than 2 million hits annually from both national and international visitors. Using stakeholder input and guidance from an editorial board, it is designed specifically to enhance information access to those who are mobile and those working at one or more sites rather than one central location. The contents of eLCOSH are searchable by trade, hazard, and type of jobsite, with a separate section on training. The site currently houses more than 800 program outputs, many of which have been translated into Spanish. In addition, there are approximately 700 links to the eLCOSH homepage and more than 14,000 links to subpages. Although the clearinghouse has served as a passive method for disseminating information, the program has actively promoted the site to stakeholders by sending them promotional fliers and posters, making presentations nationally and internationally to union and non-union leaders, safety and health professionals, and trainers. The program also developed and ran English and Spanish public service radio announcements in four major metropolitan cities and in some smaller markets in southwestern states.

The final passive R2P technique is the Construction Solutions database. This is still being developed with stakeholder input and will contain answers and provide solutions to specific questions about construction safety and health. It will also be similar to eLCOSH with respect to the target audience (i.e., mobile and dispersed workers, contractors, and supervisors working at construction sites). An interactive mechanism will allow users to provide feedback and additional information to the database about the solutions on the basis of their own personal experience.

The program has several active R2P methods, which include the following: (1) presenting findings at national and international construction safety and health and scientific conferences, (2) hosting and sponsoring conferences to engage partners across the industry, (3) publishing articles in peer-reviewed journals and relevant trade publications, and (4) using social marketing methods. The fourth method is a fairly new approach for conducting strategic dissemination to identified target populations.

OVERALL EVALUATION OF RELEVANCE

The charge to the committee required an overall assessment of the relevance of the Construction Research Program's activities to the improvement of the health and safety of construction workers. The Framework Document (Appendix A) ties relevance scores to the priority of the research areas focused on and to the level of activity for transferring research to practice (Box 3.4).

The scoring criteria for relevance are tied to the priority of the research areas focused on and to the level of activity for transferring research to practice. The committee found that the program's priorities for safety-related goals were closely aligned with national and state surveillance data identifying the leading causes of fatalities and injuries. Similarly, its focus on Hispanic workers, the largest "special population" within construction, was appropriate and of high priority.

The program's process for prioritizing research on health hazards was not as transparent. This can be attributed in part to the lack of national and state surveillance data regarding the extent of health hazards caused by specific agents and in

BOX 3.4
Framework Document Scoring Criteria for Relevance

5 = Research is in high-priority subject areas and NIOSH is significantly engaged in appropriate transfer activities for completed research projects/reported research results.

4 = Research is in priority subject areas and NIOSH is engaged in appropriate transfer activities for completed research projects/reported research results.

3 = Research is in high priority or priority subject areas, but NIOSH is not engaged in appropriate transfer activities; or research focuses on lesser priorities but NIOSH is engaged in appropriate transfer activities.

2 = Research program is focused on lesser priorities and NIOSH is not engaged in or planning some appropriate transfer activities.

1 = Research program is not focused on priorities and NIOSH is not engaged in transfer activities.

SOURCE: Reprinted from Box 2 of "Framework for the Review of Research Programs of the National Institute for Occupational Safety and Health," reproduced as Appendix A in this report.

comparison with the health hazards caused by other agents. However, it was clear that the research areas chosen for health-related hazards did affect large numbers of workers across the entire construction industry.

The committee discussed each of the four program goals and their associated research activities at length. The discussion involved a very deliberate process of examining the language for the scoring criteria. Immediate consensus emerged that at a minimum a rating of 4 would apply for each of the goals—that is, research is in priority areas and the program is engaged in appropriate transfer activities. The discussion then turned to a closer examination of the criteria for a score of 5 to determine whether the research conducted was in "high-priority" subject areas and whether the program was "significantly engaged in appropriate transfer activities for completed research projects/reported research results." (The scoring guidance does not allow for the assignment of scores using decimals, for example, 4.5).

The committee determined that the Construction Research Program was clearly engaged in high-priority activities given its focus on the leading causes of fatalities (Goal 1), health hazards that affect large numbers of construction workers (Goals 2 and 3), and significant special populations (Sub-goal 4.2).

The committee also determined that the Construction Research Program was significantly engaged in appropriate transfer activities. Across the program, R2P activities, previously described in detail, have involved a wide range of industry stakeholders, technologies, training methods, and information-dissemination activities. The program has contributed to the development of OSHA standards and worked with state agencies and industry stakeholders to transfer information and protective measures directly to the worker in the field. Stakeholder groups indicated to the committee that program-generated publications brought value to the industry by offering a means for informing their management, staffs, and members about newly developed or improved industry practices. During its deliberations, the committee also considered external factors, specifically the level of resources available to the program.

Given the committee's determination that the research conducted was high priority in nature and that the program was significantly engaged in appropriate transfer activities, the committee assigned the Construction Research Program a score of 5 for relevance.

The committee did not view this score as a statement that the program could not be improved, however. The high score instead reflects the guidance for ranking established in the Framework Document and the committee's recognition of the financial constraints within which the program has operated. For these reasons, the committee's evaluation is retrospective. During the course of its discussions and evaluation, the committee identified several areas on which the program should focus in order to maintain its excellence and continue its work in high-priority areas. These are discussed in Chapter 4.

ASSESSMENT OF IMPACT

To determine the impact that the NIOSH Construction Research Program has had on reducing safety and health hazards in construction projects, the committee focused on the intermediate and end outcomes of the program as outlined in the Framework Document. These outcomes are defined as follows:

- **Intermediate outcomes:** Related to the program's association with behaviors and changes at individual, group, and organizational levels in the workplace. An assessment of the worth of NIOSH research and its products by outside stakeholders (e.g., production of standards or regulations based in whole or in part on NIOSH research; attendance at training and education programs sponsored by other organizations; use of publications, technologies, methods, or recommendations by workers, industry, and occupational safety and health professionals in the field; and citations of NIOSH research by industry and academic scientists).
- **End outcomes:** Improvements in safety and health in the workplace. Defined by measures of health and safety and of impact on processes and programs (e.g., changes related to health, including decreases in injuries, illnesses, or deaths and decreases in exposures due to research in a specific program or subprogram).

The committee recognized first that the program cannot directly impose change in the workplace or change in workplace behaviors that is necessary to reduce fatalities, serious injuries, and work-related illnesses. What the program can be expected to do is to contribute evidence-based knowledge about hazards and the prevention of adverse outcomes. The program can also be expected to promote plausible, evidence-based, risk-reducing actions by others, including OSHA and other regulatory agencies, equipment manufacturers, construction project owners, contractors, and workers. For this reason, the committee primarily relied on intermediate outcomes in its evaluation of impact. The committee also considered external factors that bear on the incorporation of program outputs in workplace practices, particularly the segmentation of the industry.

Goal 1: Reduce the Major Risks Associated with Traumatic Injuries and Fatalities in Construction

The reduction of fatalities and serious injuries within the construction industry is a significant end outcome for any safety-related program. During the period reviewed, 1996 through 2005, the rate of fatalities in the construction industry declined by 3.2 deaths per 100,000 workers. Given a workforce of 8 million, this translates to about 250 fewer work-related deaths per year. The rate of injuries also declined significantly, perhaps by as much as 45 percent, although changes in statistical data-gathering methods make it difficult to compare data over the entire review period.

The question before the committee was the extent to which the Construction Research Program and its activities contributed to these positive outcomes. The answer was not clearcut. A number of factors have been cited for these declines, including a greater emphasis on safety by owners and contractors in the industrial and infrastructure sectors of the industry and by some unions. Additional factors include increasing health care costs, which create a greater awareness of the costs of unsafe practices, and business school curriculums that stress the significance of management commitment in the implementation of safety programs. All of these represent management and behavior/awareness practices.

The Construction Research Program has primarily focused on technologies for safety, in areas including fall protection and interventions, contact with electricity, proximity-warning systems, trenching and excavation standards, and training activities.

The program has also generated a range of intermediate outcomes that have likely helped to reduce traumatic injuries. Intermediate outcomes have included hundreds of publications and presentations, training, and training aids. Research generated by the program has been incorporated in standards and course curricula used by some states and industry associations and in OSHA standards for working in confined spaces. Although it is not possible to determine exactly how much the program has contributed to the decline in fatality and injury rates on construction sites, the committee concluded that the program did have some impact on both end outcomes and well-accepted intermediate outcomes.

Goal 2: Reduce Exposures to Health Hazards Associated with Major Risks of Occupational Illness in Construction

Data that could be used to measure end outcomes, such as decreases in the number of cases of silicosis, hearing loss, or lead poisoning, are not available. For that reason, the committee's evaluation of the program's impact for the Goal 2 area focused almost exclusively on intermediate outcomes. In the area of hearing loss, program research has been important in the development of an evidence-based, revised hearing-conservation standard for construction that could be issued by OSHA in the future, as well as a new American National Standards Institute (ANSI) standard for preventing hearing loss in construction. Other training materials and methods developed by the program are being adopted by the Carpenter's Union for its 500,000 members and by construction contractors in Washington State.

In regard to lead exposures, New York and New Jersey have used the Construction Research Program work on surveillance of blood lead levels to enhance their focused efforts to address construction worker exposures in their states, including the development of outreach and educational materials for use in high-risk construction activities. Surveillance data have also been used to evaluate the effectiveness of intervention measures. In addition, ABLES data have been used by OSHA in performing the review of its construction lead standard. Data and guidance developed by the Construction Research Program have also been used by the Steel Structure Painting Council's painting contractor certification program that assesses a contractor's ability to protect worker health and safety on projects involving lead coatings on steel.

The field-portable techniques for sampling and analysis developed by the program were used by the EPA in developing its regulations on renovation and remodeling for residential abatement projects.

The program's lead "take-home" work was used by Congress in developing legislation requiring HUD to conduct a lead-paint-abatement demonstration program. At HUD's request, the program evaluated the worker protection measures in its demonstration projects. The report to Congress on take-home issues has also been cited as reference material by the National Association of the Remodeling Industry. Other exposure characterization data gathered by the Construction Research Program have been used by federal government agencies, including HUD and OSHA, in the development of their requirements and guidance for worker protection. Several states, including California and Massachusetts, used the exposure characterization data to develop their programs to reduce lead exposures for construction workers.

The model specifications for worker protection measures on steel structures, developed by the program, have been used by the Federal Highway Administration and a number of states (New York, New Jersey, Michigan, Maryland, Missouri) to assist them in developing job specifications for such work in their states.

The Construction Research Program has also generated a range of intermediate outcomes regarding silica and silicosis. The program's risk-assessment characterizations and hazard-review document for silica have been used by several organizations, including the California Office of Environmental Health Hazard Assessment and the United Kingdom's Health and Safety Executive, in setting standards.

The work conducted on engineering control development for a variety of tasks generating airborne silica (e.g., working with masonry products or using jackhammers) has been adopted by some organizations, including trade organizations, labor unions, and contractor associations.

Information on control measures generated by the program has also played a role in the promulgation or development of standards designed to reduce silica exposures in construction. California OSHA has used data from the program to regulate silica exposures, and New Jersey used data to pass legislation prohibiting the dry grinding and sawing of masonry products. Likewise, OSHA has used the engineering control information generated by the program for preparing its cost and technological feasibility analysis as part of its work on a proposed silica standard for construction. All of these government regulatory efforts are likely to reduce exposures and protect workers from the hazards of silica.

The analytical work conducted by the Construction Research Program has resulted in the lowering of the limits at which airborne concentrations of silica can be accurately and reliably measured. This major advancement will help set the stage for lowering the OSHA permissible exposure limit to a level that can now be reliably measured. The lowering of the exposure limit should in turn reduce the risk of workers for developing silicosis and lung cancer from inhaling respirable silica.

Construction workers operating heavy equipment in cabs are exposed to airborne silica and are at risk of developing silicosis. In response to work with the Construction Research Program, a cab filtration system has been commercially developed, as has a patented leak test for use with cabs. As a result, workers in cabs with the filtration equipment are likely to have reduced exposures to airborne silica. Overall the silica exposure control program has generated a substantial number of intermediate outcomes that could result in fewer new cases of silicosis among construction workers.

A rare example of the achievement of an end outcome for health is the program research with commercial asphalt pavers and the development of engineering controls. The program has demonstrated that engineering controls reduce asphalt fume exposure by 50 to 80 percent (Mickelson et al., 1999, 2006). Because the controls have been broadly implemented and have been well accepted within the highway-paving industry, asphalt fume exposures among 300,000 highway-class road-paving workers have been significantly reduced. In addition, evaluations of control on new pavers at 12 sites across the United States have indicated that workers' personal exposures to total particulate matter and benzene-soluble matter (two primary contaminants of concern) were both consistently reduced to levels below U.S. government-recommended values (Mickelson et al., 2006). This project represented a good example of how the NIOSH research process has benefited the health and safety of both unionized and non-unionized construction workers. This effort on behalf of the pavement workers also generated more attention to asphalt exposure control for roofing industry workers.

In contrast to these activities, the program's work in the area of dermal exposures has been limited, due in part to the narrow focus on skin disorders associated with the use of Portland cement.

Regarding welding fume exposures, the Construction Research Program has produced several important intermediate outcomes. Program evaluations of the health-effects literature have been used by the American Welding Society in its assessments, and OSHA used, at least in part, the cancer risk assessments from the program for reducing the exposure limit for hexavalent chromium. The new OSHA hexavalent chromium standard and a reduced exposure limit will potentially impact all welders across the United States, including those in the construction industry.

A more limited intermediate outcome has occurred in relation to the program's work on exposure characterization and intervention research. Local exhaust-ventilation systems have been put in place on a major power plant turn-around project, and several pipe fitter apprenticeship programs have upgraded their local exhaust-ventilation systems at their training centers.

Goal 3: Reduce the Major Risks Associated with Musculoskeletal Disorders in Construction

Research performed or funded by the Construction Research Program has been of critical importance to other researchers focusing on musculoskeletal disorders. Half of published research on MSDs in construction has been performed or funded by the program. It is safe to say that over the past decade there are virtually no significant studies of MSDs in construction that have not cited NIOSH-supported work. An important output from this research is the building of additional research and training capacity at the universities that have received NIOSH funding. Program funding not only produces research aimed at measuring and reducing MSDs and risk factors for MSDs, but also provides the means to educate future practitioners and researchers.

A review of intermediate outcomes and their impact shows a number of notable accomplishments. Most striking is that from 1990 to 2003, the NCC researchers were responsible for 50 percent of all journal publications about ergonomic hazards and controls in the U.S. construction industry. A review of the current construction literature since 2003 reveals that a similarly high proportion of peer-reviewed journal articles assessing MSDs in construction has been done by researchers in the Construction Research Program, through program extramural grants, or in collaboration with other NIOSH researchers.

Key program publications on construction ergonomics and the design of hand tools have been widely distributed, with 130,000 hard copies distributed and almost 200 downloads per month from the NIOSH Web site. In surveillance activities, there have been more than 25 journal articles and 9 conference presentations on exposure assessment, 63 journal articles and 30 presentations on MSDs, and

30 journal articles or presentations, 28 newsletter articles, and 3 NIOSH publications on intervention evaluation (NIOSH, 2007). The large number of practice-oriented materials disseminated indicates significant interest in program-generated research by people in the health and safety community.

A number of research tools created by the program have been used by others in the health and safety research community. A questionnaire for surveillance of MSDs among construction workers has been translated into several languages and used by other researchers. The portable data logger for exposure assessment and also the Posture, Activity, Tools, and Handling (PATH) observational exposure-assessment method for construction are being used by other research teams outside NIOSH. In an example of crosscutting research and technology transfer, the data logger and PATH method have both been adapted for use in other industries.

A number of training programs have been based on Construction Research Program materials. Using program funding, the United Brotherhood of Carpenters developed an ergonomics training program that has trained more than 60 instructors and hundreds of carpenters in ergonomics related to construction work. The Smart Mark module is used nationally, with more than 4,000 instructors training more than 50,000 construction workers annually. This health and safety program includes an ergonomics module directed at MSDs; other portions of the Smart Mark program are directed at avoiding common musculoskeletal acute injuries such as sprains and strains. Other interventions have included the identification of overhead drilling as a hazard. These interventions have been approached both through participatory ergonomics, which resulted in changing practices during the "Big Dig" construction project in Boston. The program has also funded a project to develop an improved overhead drilling method using a drill jig. This method was developed by the University of California and has been demonstrated at the Construction Safety Council meeting as well as at other venues. Plans to commercialize this innovation are under way.

Interventions have also included evaluations of the effectiveness of participatory ergonomics programs on the incidence, severity, and cost of MSDs on a large construction project. The program has been active in the design of hand tools. Sheet metal workers were previously identified as a group with a very high prevalence of carpal tunnel syndrome. Using several exposure-assessment methods, program researchers compared the musculoskeletal load on the wrists and forearm using conventional metal snips. They worked with a major manufacturer of hand tools to develop a new metal snip that is less stressful to the hand and wrist. These snips are now commercialized and selling well in a variety of retail outlets carrying hand tools.

A participatory ergonomics program on the incidence, severity, and cost of MSDs on a large construction project demonstrated a 25 to 35 percent cost reduction at sites using the participatory ergonomics program compared with costs at other sites not using the program (NIOSH, 2007). Intel Corporation has now adopted this program for use on all its international and U.S.-based construction projects.

Other translation and dissemination activities have included the Oregon Construction Ergonomics Initiative (CEI), a collaborative group of industry professionals including contractors and labor representatives and supported by Oregon OSHA. In addition to disseminating information, the CEI has worked with Oregon OSHA to provide consultation to a wide range of construction companies.

The program publication *A Guide to Selecting Non-Powered Hand Tools* has had more than 100 citations or links by organizations in the United States, Canada, Brazil, Japan, and Europe (NIOSH, 2007). This publication is an example of the impact of program education dissemination on a variety of audiences, including health and safety professional associations, government agencies, educational institutions, trade unions, trade associations, and health and safety resource guides.

Major new technology and method developments from the program's vibration research include instrumentation of handles and biodynamic response measurement methods that are used by researchers all over the world: for example, development of new tool- and glove-testing methods, development of a new method to characterize the grip force applied on cylindrical handles, and a novel three-dimensional hands-on vibration test system that has become a commercial product in a joint effort with two private companies. Other outcomes have included major contributions to International Organization for Standardization and ANSI guidelines on vibration.

A major factor limiting the impact of the program's research on health and safety is the absence of regulatory requirements to prevent work-related MSDs, except for the relatively weak standard that exists in the state of California. OSHA promulgated ergonomics regulations late in 2000 based in large part on research performed under program leadership. This standard was struck down in 2001. The Washington State ergonomics regulation, which had a large construction component, was struck down in a referendum in 2004. The absence of a specific standard for the prevention of work-related MSDs has made it more difficult for research findings to be translated into changes in equipment or behavior.

Goal 4: Increase Understanding of Construction Sector Attributes That Affect Occupational Safety and Health Outcomes

Assessing the impact of Goal 4 activities presented a different challenge from that of assessing the impact of Goals 1 through 3, given the nature of the research goals of Goal 4.

An effective resource for surveillance data is *The Construction Chart Book: The U.S. Construction Industry and Its Workers* (CPWR, 2007). This source provides a compilation of data that illustrates how the industry is performing, growing, and changing. Its fourth edition was published in December 2007. Among its many illustrations and data charts, the book also addresses the surging Hispanic population in the construction industry.

Although the Construction Research Program has made efforts to produce outputs in the form of publications targeting small businesses, and although many of these publications have been produced in Spanish, the challenge remains about how best to get this information to the target audiences. The program has taken important first steps in addressing the safety and health of the Hispanic segment of the U.S. construction workforce despite the complexity and challenging nature of the problem. These steps include collaboration with NIOSH's Traumatic Injuries Program, the NIOSH Occupational Health Disparities Coordinated Emphasis Area, community and day-labor organizations, and colleagues within state and federal government. Key focus areas for program researchers have included the identification of major socioeconomic and work organization factors contributing to health and safety disparities, along with the impact of language differences.

With respect to the program's R2P efforts, a number of barriers still exist. First, extramural projects funded by the program were directed to use 20 percent of direct costs for R2P efforts. However, given the size, diversity, and segmentation of the construction industry, the aforementioned funding level has been insufficient to implement more active R2P dissemination strategies and to evaluate their effectiveness. Outputs generated by external grantees and/or partners should be included by the program's researchers in the program's R2P efforts as well.

Internally, program researchers have been encouraged to translate research findings to lay publications for target audiences and stakeholders. Indeed, an R2P plan is now required for all internally funded projects. However, this activity falls outside the mainstream job duties for the NIOSH bench scientists, chemists, engineers, epidemiologists, and industrial hygienists. In addition, many do not have the expertise for preparing or producing such documents other than those for peer-reviewed journals.

OVERALL EVALUATION OF IMPACT

In addition to its other assessment tasks, the charge to the committee requires an overall assessment of the impact of the Construction Research Program. The scoring criteria are summarized in Box 3.5.

The committee evaluated the impact of the program using the same process that it used for relevance. The scoring criteria for impact are linked to a program's

BOX 3.5
Framework Document Scoring Criteria for Impact

5 = Research program has made major contribution(s) to worker health and safety on the basis of end outcomes or well-accepted intermediate outcomes.

4 = Research program has made some contributions to end outcomes or well-accepted intermediate outcomes.

3 = Research program activities are ongoing and outputs are produced that are likely to result in improvements in worker health and safety (with explanation of why not rated higher). Well accepted outcomes have not been recorded.

2 = Research program activities are ongoing and outputs are produced that may result in new knowledge or technology, but only limited application is expected. Well accepted outcomes have not been recorded.

1 = Research activities and outputs do not result in or are NOT likely to have any application.

SOURCE: Reprinted from Box 3 of "Framework for the Review of Research Programs of the National Institute for Occupational Safety and Health," reproduced as Appendix A in this report.

contributions to worker health and safety based on end outcomes or well-accepted intermediate outcomes. In terms of end outcomes, the committee concluded that the Construction Research Program, through its development of some technologies such as fall-protection equipment and proximity-warning systems, has made some contributions to the overall declines in fatalities and injuries, although the full extent of that impact is not known. Additionally, the program has had a positive impact on the health of workers exposed to asphalt fumes generated during road-paving operations.

The program has also been responsible for a large range of intermediate outcomes. Its research on musculoskeletal disorders is cited in about half of all publications on this topic (NIOSH, 2007). The program has provided evidence for the development of OSHA standards on ergonomics, hearing conservation, respiratory crystalline silica, trenching practices, and lead in construction. Some of these standards have been issued, others have not. However, whether the standards are issued and enforced is beyond the control of the program. Its training and training dissemination activities have been extensive, and it is likely that they have contributed to the prevention and reduction of health and safety hazards on some construction worksites.

The committee also determined that the segmentation of the industry and the less-than-adequate level of resources have had a bearing on the program's impact. Thus, although program-generated publications, technologies, and training are relevant for all segments of the construction industry, their diffusion has varied by construction sector. It is particularly difficult to reach the residential sector because so many residential contractors are self-employed or employ fewer than 10 workers. The level of funding available limits the ability of the program to conduct surveillance research and to provide more direct training to owners and workers in this sector.

Using the scoring criteria for rating the program's impact, the committee determined that the Construction Research Program has made *some* contributions to construction health and safety as measured by either end outcomes or well-accepted intermediate outcomes. However, committee members had divergent views about whether these contributions could be classified as *major* contributions across the entire program. On that basis, the committee assigned the program an impact score of 4. As it did with respect to the program's relevance, the committee made recommendations intended to improve the program's impact in the future. These are presented in Chapter 4.

REFERENCES

BLS (Bureau of Labor Statistics). 2007. Occupational Injuries, Illnesses, and Fatalities. Available at http://www.bls.gov/iif/#tables. Accessed August 15, 2008.

Cawley, J.C., and G.T. Homce. 2006. Occupational electrical injuries in the United States, 1992-2002. Pp. 325-338 in Proceedings of the IEEE Petroleum and Chemical Industry Committee Annual Conference, September 11-13, 2006, Philadelphia, Pa.

CDC (Centers for Disease Control and Prevention). 2007. Nail gun injuries treated in emergency departments—United States 2001-2005. In Morbidity and Mortality Weekly Report 56(14):329-332.

CDC. 2008. Toxicological Profile Information Sheet. Available at http://www.atsdr.cdc.gov/toxprofiles/tp100-c9.pdf. Accessed August 15, 2008.

CPWR. 1999. Employers Guide to Skin Protection for Work with Wet Cement in Construction. Available at http://www.cdc.gov/elcosh/docs/d0400/d000457/d000457.html.

CPWR. 2007. The Construction Chart Book: The U.S. Construction Industry and Its Workers. 4th ed. Silver Spring, Md.: Center for Construction Research and Training.

Dement, J., K. Ringen, L. Welch, E. Bingham, and P. Quinn. 2005. Surveillance of hearing loss among construction and trades workers at Department of Energy nuclear sites. American Journal of Industrial Medicine 48:348-358.

Espinosa, A.J., J.J. Sanchez Hernandez, F.P. Bravo, M. Gonzalez-Baron, A.P. Zamora, and A.E. Espinosa. 1999. Cutaneous malignant melanoma and sun exposure in Spain. Melanoma Research 9(2):199-205.

IOM (Institute of Medicine). 2006. Hearing Loss Research at NIOSH. Washington, D.C.: The National Academies Press.

Meyer, S. 2003. Fatal occupational injuries involving confined spaces. Occupational Safety and Health 72(11):58-64.

Mickelson, R.L., K.R. Mead, S.A. Shulman, and T.E. Brumagin. 1999. Evaluating engineering controls during asphalt paving using a portable tracer gas method. American Journal of Industrial Medicine Supplement 1:77-79.

Mickelson, R.L., S.A. Shulman., A.J. Kriech, L.V. Osborn, and A.P. Redman. 2006. Status of worker exposure to asphalt paving fumes with the use of engineering controls. Environmental Science and Technology 40(18):5661-5667.

MMWR (Morbidity and Mortality Weekly Report). 2004. Occupational fatalities during trenching and excavation work—United States, 1992-2001. MMWR 53(15):311-314.

NIOSH (National Institute for Occupational Safety and Health). 2007. NIOSH Construction Research Program Evidence Package. Washington, D.C., July.

NRC (National Research Council). 2001. Musculoskeletal Disorders and the Workplace: Low Back and Upper Extremities. Washington, D.C.: National Academy Press.

Society of Toxicologic Pathology. 2006. Welding fume exposure and associated inflammatory and hyperplastic changes in the lungs of tumor susceptible mice. Toxicolology Patholology 34:364.

4

Future Research and Program Improvement

The committee was asked to address the Construction Research Program's effectiveness in targeting new research areas and identifying emerging issues most relevant to future improvements in the health and safety of workers in construction workplaces. The committee was also asked to make recommendations to improve the program. Because there is some overlap among these tasks, the committee chose to discuss all of these issues together in one chapter.

The first section of Chapter 4 provides a brief overview of the Construction Research Program's new and emerging research areas as presented in the evidence package (NIOSH, 2007) provided to the committee by the program staff of the National Institute for Occupational Safety and Health (NIOSH). The next section, in accord with the Framework Document (Appendix A), presents the committee's analysis of and recommendations regarding emerging issues and new research areas. The analysis and recommendations are based on discussions with stakeholders and on individual committee members' knowledge of the construction industry and their backgrounds and expertise. The committee also considered the resources available to the program. Thus, although it recognized that nanoparticles and nanotechnology are emerging as areas of research, the committee believed other areas to be of higher priority. Chapter 4 concludes with the committee's recommendations for overall program improvement.

OVERVIEW OF NEW AND EMERGING RESEARCH AREAS

The Construction Research Program's research agenda has evolved over time on the basis of identified health and safety needs in the construction industry:

- *Phase 1: 1990-1995*—The research agenda focuses on needs assessment and developing surveillance capacity.
- *Phase 2: 1996-2004*—The first National Occupational Research Agenda (NORA1) focuses on risk-specific intervention research, including a special focus on musculoskeletal disorders and ergonomics.
- *Phase 3: 2005 and beyond*—NORA2 focuses on translation and diffusion of research (research-to-practice).

During each of the three phases, multiple methods were used to identify new and emerging issues. The next phase of priority setting will be focused on the second iteration of the National Occupational Research Agenda. NORA2 differs from NORA1 in being sector-based; one of the sectors is construction.[1] The intent is to expedite the translation of research into practice in the workplace. The NORA Construction Sector Council will play a key role in bringing together researchers and practitioners to identify and prioritize the challenges and related research needs facing the construction industry now and in the future. The Construction Sector Council will also need to coordinate with other sector councils to explore and prioritize some crosscutting topics, for example, that of workers struck by vehicles or equipment on road projects, which affects both the construction and transportation sectors.

In the evidence package provided to the committee (NIOSH, 2007), the NIOSH program staff noted that many of the predominant characteristics of the construction industry that currently have an impact on health and safety on the job site (e.g., short-term contracting, the predominance of small employers with high turnover among their employees, temporary employment, multiple-employer worksites, a multicultural workforce, and episodic exposure to risks) will not change significantly. The staff also identified a number of anticipated changes that may have safety and health implications for the future. These changes, which may warrant targeted research activities, include the likelihood that financing costs, project management costs, and costs of supplies, technology, equipment, and energy will rise. Such cost increases may lead construction company owners in some segments of the industry to pursue cost savings by hiring more unskilled, lower-paid workers; engaging in subcontracting rather than direct hiring of prime contractors; and/or

[1]The sectors are Agriculture, Forestry and Fishing, Construction, Healthcare and Social Assistance, Manufacturing, Mining, Services, Transportation, Warehousing and Utilities, and Wholesale and Retail Trade.

placing an increased demand on construction labor productivity, including pressure on construction schedules. The evidence package states that this desire of construction company owners for increased productivity may lead to their exerting pressure on design and engineering firms to develop new (and untested) construction methods and to putting pressure on production labor to work for lower wages or to produce more per unit cost.

In reviewing the evidence package, the committee determined that the Construction Research Program's process for setting priorities for conducting future research on health hazards is not clear or evident. For example, there is little emphasis and limited focus on dermal exposure in construction, yet it remains a significant route of exposure to hazardous chemicals in this industry. A transparent process governing the resource and time commitments to be used in selecting high-priority health-related projects would assist the program and its stakeholders in their decision-making processes.

Priority Topics for Future Research

In 2002, the Construction Sector Council, which provides direction for the Construction Research Program's activities, identified a number of priority topics in which research would be most likely to improve the program's impact. These topics were grouped into three categories:

1. Health and injury outcome topics that target the following:
   - Leading types of fatal and nonfatal traumatic injuries in construction;
   - Low-back injuries and other cumulative work-related musculoskeletal disorders among construction workers; and
   - Occupational illness topics that focus on respiratory disease and hearing loss. Respiratory disease includes airways disease, asthma, chronic obstructive lung disease, and silicosis.
2. Chemical and physical exposure topics that target the following:
   - Vibration,
   - Asphalt fumes, and
   - Lead and dust particles.
3. Approaches and sector topics that target the following groups and issues within construction:
   - Small and self-employed contractors;
   - Special subpopulations at risk within construction, such as Hispanic workers, day laborers, young workers, and aging workers;
   - The role of project design as a primary prevention tool for addressing construction hazards;

- Addressing work organization in construction and improving the understanding of how it affects health and safety;
- Working with building owners and clients to promote and evaluate construction best practices; and
- Leveraging promising approaches from related high-risk sectors such as agriculture and mining into construction.

Some of these topics were further developed into NIOSH Construction Research Program strategic goals in 2005. NIOSH shared these strategic goals as input to the NORA Construction Sector Council in 2006, and most but not all of these 2002 and 2005 topics were subsequently incorporated in some form into the NORA2 Preliminary Draft National Construction Agenda Strategic Goals (Box 4.1). The draft National Construction Agenda is the first national effort to create an agenda for health and safety in the construction industry. Goals 1.0 through 7.0 are classified as "outcome goals" that will result in actual reductions in injuries, exposures, illnesses, and disorders among construction workers. Goals 8.0 through 14.0 are classified as "contributing-factor goals." These goals are defined as factors that represent important influences impacting the likelihood that prevention and control measures and actions are taken on a construction site.

For each of the 14 strategic goals, the NORA2 agenda includes intermediate goals that reflect an intermediate step and outcome necessary to move toward a strategic goal. In some cases, contributing-factor goals overlap with outcome goals. For example, an intermediate goal for preventing falls to lower levels at a construction site (Goal 1.0) is to partner with architects, engineers, and construction organizations to expand the use of prevention-through-design practices (Goal 13.0).

Some of the topics in the categories above represent areas in which the program is already engaged. These areas offer important opportunities to move research to practice (R2P). Other topics include some new areas for research.

In the following section, the committee's recommendations for future research areas are, to the extent possible, presented within the context of the NORA2 goals. The committee is particularly interested in emphasizing to the Construction Research Program staff that it place an increased amount of research time, effort, and resources on the contributing-factors goals—specifically, Strategic Goals 8.0 through 14.0.

For all of the goals, the committee also recommends that the program keep the worker and contractor in mind as the ultimate destination for its R2P efforts. Following are two critical research questions that should remain in the forefront of these efforts: (1) How can the program get vital information to the worker "in the trench" or "on the steel"? and (2) How does the program persuade contractors and workers to effectively use the interventions that are developed through research?

BOX 4.1
NORA2 Preliminary Draft National Construction Agenda Strategic Goals

STRATEGIC GOAL 1.0—Reduce construction worker fatalities and serious injuries caused by falls to a lower level.

STRATEGIC GOAL 2.0—Reduce fatal and nonfatal injuries from contact with electricity among construction workers.

STRATEGIC GOAL 3.0—Reduce fatal and serious injuries associated with struck-by incidents associated with objects, vehicles, and collapsing materials and structures.

STRATEGIC GOAL 4.0—Reduce hearing loss among construction workers by increased use of noise reduction solutions, practices, and hearing conservation programs by the construction community.

STRATEGIC GOAL 5.0—Reduce silica exposures and future silicosis risks among construction workers by increasing the availability and use of silica dust controls and practices for tasks associated with important exposures.

STRATEGIC GOAL 6.0—Reduce welding fume exposures and future related health risks among construction workers by increasing the availability and use of welding fume controls and practices for welding tasks.

STRATEGIC GOAL 7.0—Reduce the incidence and severity of work-related musculoskeletal disorders among construction workers in the U.S.

STRATEGIC GOAL 8.0—Increase understanding of factors that comprise both positive and negative construction safety and health cultures; and, expand the availability and use of effective interventions to maintain safe work practices 100% of the time in the construction industry.

ANALYSIS OF AND RECOMMENDATIONS REGARDING EMERGING ISSUES AND NEW RESEARCH AREAS

Reduce Fatalities and Nonfatal Injuries (NORA2 Goals 1.0, 2.0, and 3.0)

The NORA2 draft agenda focuses on workers' falls to lower levels at construction sites, their contact with electricity, and their being struck by objects, vehicles, and collapsing materials and structures. The Construction Research Program has already conducted relevant and valuable research in reducing fatalities and nonfatal

STRATEGIC GOAL 9.0—Improve the effectiveness of safety and health management programs in construction and increase their use in the industry.

STRATEGIC GOAL 10.0—Improve understanding of how construction industry organization factors relate to injury and illness outcomes; and increase the sharing and use of industry-wide practices, policies, and partnerships that improve safety and health performance.

STRATEGIC GOAL 11.0—Increase the recognition and awareness of construction hazards and the means for controlling them through broad dissemination of quality training for construction workers, including non-English speaking workers.

STRATEGIC GOAL 12.0—Increase understanding of how vulnerable worker groups experience disproportionate risks in construction work and expand the availability and use of effective interventions to reduce injuries and illnesses among these groups.

STRATEGIC GOAL 13.0—Increase the use of "prevention through design (PtD)" approaches to prevent or reduce safety and health hazards in construction.

STRATEGIC GOAL 14.0—Improve surveillance at the Federal, State, and private level to support the identification of hazards and associated illnesses and injuries; the evaluation of intervention and organizational program effectiveness; and the identification of emerging health and safety priorities in construction.

SOURCE: Reprinted from NORA Construction Sector Council (2008).

injuries from workers' falls and contact with electricity. In some areas where the knowledge base has been well developed, such as in the avoidance of the use of aluminum ladders when doing electrical work, or in improving safety performance through routine inspections, future activities should focus on R2P. The committee supports a shift in research to less developed areas such as incidents in which workers are struck by vehicles, equipment, or objects.

The committee cautions, however, that although research on fatalities from these sources is expected to result in safety improvements with respect to nonfatal traumatic injuries as well, such a fatality-driven approach may not adequately address other important areas for nonfatal injuries. One such area has been falls occurring at the same level. Such slip, trip, and fall injuries have been a major source of morbidity and disability in general industry and are likely to represent a major problem in construction as well. Since these injuries infrequently result in fatalities, they have not been captured in the current priority-setting system. Capturing and tracking the downstream effects of such nonfatal injuries would provide a clear picture of the types of upstream research that could serve to reduce such injuries. Research in this area should address methodologies for identifying and including other causes of traumatic injury, along with how to improve dissemination of the research results, particularly to vulnerable populations.

The committee believes that "frequency of injuries" should be considered as a criterion for setting priorities with regard to research topics addressing traumatic injury. Other factors for consideration could be the ease of implementation of evidence-based interventions or the cost of implementation.

Much of the Construction Research Program's work has involved technological or engineering solutions to safety issues—for example, fall-protection equipment or warning systems regarding proximity to overhead power lines. What appears to be needed is a greater focus on the use of these technologies in the field. An example is equipment that has been developed for fall protection. During the stakeholder panels conducted by the committee, some stakeholders indicated that wearing this equipment is more dangerous than not wearing it. This problem indicates to some degree that the program is failing to effectively disseminate to safety trainers and stakeholders the information pertaining to the proper use of fall-protection gear. The existing mind-set within the industry that fall-protection gear is not valuable, or even worse that it is perceived to be dangerous, underscores a disconnect between the practice of using this gear and the true benefits and protection that can be derived from using it. The committee suggests that future research should focus more on barriers to implementation of fall-protection equipment or other technologies. A potentially fruitful area of research collaboration that could perhaps hasten the adoption of innovative solutions would be with construction equipment manufacturers.

In the committee's discussions with stakeholders on injury and fatality risk reduction, stakeholders expressed interest in using excavation as an issue to learn more about implementation issues. The Occupational Safety and Health Administration (OSHA) trenching standard approach is viewed as effective, and the remaining obstacles to prevention of injuries from collapsing materials appear to be related to raising awareness and getting contractors and workers to use existing standards. This issue represents a potential new R2P research area, with stakeholder interest in the continuation of efforts to explore other approaches of OSHA's Advisory Committee for Construction Safety and Health trenching work group and by members of the NORA Construction Sector Council.

Reduce Hearing Loss (NORA2 Goal 4.0)

It is well established that exposure to noise causes hearing loss and that prevention programs are effective in protecting the hearing of construction workers. Reducing construction workers' exposure to causes of hearing loss and reducing the numbers and rate of hearing-loss cases ought to be a principal health and safety objective of the construction industry. The primary objective of the Construction Research Program should be to develop engineering controls and to find ways to broadly implement effective hearing-loss prevention measures. The activities proposed by the Construction Research Program in the evidence package appear to be well suited to meeting this objective.

The Construction Research Program's development of models for "portable" hearing-conservation programs in construction will help address issues created by the temporary and mobile nature of the workforce. Assessing the effectiveness of implemented hearing-conservation programs will also be an important activity for identifying factors that could lead to more widespread adoption of hearing-conservation practices by contractors. Expanding the powered-hand-tool database and working with stakeholders to develop engineering control measures for hand tools will be a valuable activity.

Additional Construction Research Program activities would be properly devoted to training and guidelines that address noise controls, impulse noise, and noise caused by machinery and equipment. Continuing the program's current work to measure impulse noise and develop effective intervention measures to protect workers from the adverse effects of impulse noise will constitute an important contribution to this area.

Regarding another major source of noise emissions, heavy equipment, the committee suggests that the program examine what research would be effective in characterizing exposures from this source and work with stakeholders (equipment manufacturers, trade associations, trade unions, and others) to develop noise engineering solutions focusing on sources of emissions from the equipment. Quieter heavy equipment could play an important role in reducing noise exposures experienced by construction workers.

Reduce Silica Exposures and Future Risks of Silicosis (NORA2 Goal 5.0)

The Construction Research Program has done exceptional work in identifying emerging issues and targeting new research on silica. Planned silica research activities include several areas that deserve increasing attention, including improvements in silicosis surveillance, development and delivery of worker training programs, expanding control technology work to focus on additional construction activities that generate airborne silica, and gathering and assessing toxicology information on substitutes for abrasive blasting.

The tearing up of old asphalt highway paving and the laying down of new asphalt can generate dust containing asphalt and silica. The program has identified a future research initiative to reduce exposures to silica and asphalt during this asphalt milling operation. Such operations involve many of the same partners that participated in the now completed road-paving project.

Reduce Welding Fume Exposures and Risks (NORA2 Goal 6.0)

The Construction Research Program has identified an extensive and well-thought-out set of emerging issues involving welding fumes that will be the target of new research initiatives. Regarding health-effects research, the program has identified a significant emerging problem concerning exposure to manganese from welding operations and the possible link to neurological effects such as Parkinson's disease. Considerable concern exists regarding this possible linkage, which is currently the subject of litigation by welders across the United States. The findings from this work will assist greatly in providing firm data on which to determine whether or not risks exist. Additionally, the National Toxicology Program will be carrying out animal studies of chronic welding exposures, using the NIOSH generation and delivery system, to look at carcinogenic and neurological effects of welding fumes. Such studies, with the assistance of the program, will be important advances in the understanding of the toxicity of welding fumes. The results of this work could help identify risk reduction needs and the potential for decreasing adverse health outcomes among welders.

The program has also outlined a plan to expand its work on developing and implementing engineering controls that are designed to reduce exposures to welding fumes. Additional exposure characterizations will be undertaken, and evaluations of the impact of local exhaust ventilation will likewise be performed. Although these additional data are valuable, they only become useful in reducing risks to welders if engineering controls are implemented in the field.

The program has also identified a number of steps that it intends to initiate to ensure that the research findings from the planned activities are disseminated. The diffusion plan includes a variety of approaches that offer the potential to expand implementation, including the development of case studies, job specifications, and collaborative programs with contractors, unions, and equipment manufacturers. If these efforts are realized, exposures and risks to welders will be decreased, constituting a major accomplishment.

Reduce the Incidence and Severity of Work-Related Musculoskeletal Disorders (NORA2 Goal 7.0)

NIOSH has clearly been the leader in musculoskeletal disorders (MSD) research in the United States. The goals, activities, and outputs pertaining to this research area have addressed issues that are highly relevant for improving the musculoskeletal health of construction workers. The program has effectively acquired a diverse spectrum of inputs. It has attempted to focus its activities in this area of construction research on the major tools used and the exposures that occur in construction work. Nonetheless, obtaining surveillance data on construction-related MSDs, including exposure assessment data, has been and continues to be a challenge owing to industry segmentation and a mobile workforce.

Additional research is needed on the economics of preventing ergonomic injury. The construction program should develop a stronger evidence base in terms of describing both the need for and the impact and return on investment from the adoption of ergonomic interventions. Results of such interventions will yield progress over time, given consistent application. Research-to-practice efforts should focus on applying such interventions through community health programs designed to instill ergonomic awareness in workers. Industrywide programs are needed to bring cross-sector support for such changes.

Measurement issues pertaining to vibration are also a key emerging research area. Future research should address and identify a pathway for a healthier environment that minimizes the effects of vibration on workers.

Increase Understanding of Construction Safety and Health Cultures and Improve the Effectiveness of Management Programs (NORA2 Goals 8.0 and 9.0)

The NORA2 agenda includes a preliminary description of the construction safety and health culture as the attitudes, values, priorities, and behaviors of management and its employees and understanding the impact of this culture on the development, implementation, performance, oversight, and enforcement of safety and health.

To date, the program has conducted little research to examine systematically the impact of culture and safety and health management systems on reducing injuries and illnesses and improving conditions in the construction industry. There is evidence, however, that a matter of critical importance in preventing injuries and fatalities on construction projects is how well participating parties (owners, contractors, contractor associations, insurance carriers, and, as appropriate, unions) work together to establish a project safety culture that involves and seeks input and buy-in from the crafts. For example, work performed by the Construction Industry Institute, an organization of owners and contractors devoted to conducting research in the construction industry, has demonstrated the positive impact of safety culture and management commitment on substantially reducing injuries and fatalities.

The committee believes that the program should devote significant resources to improving the understanding of the elements of construction culture and safety and health management systems that are effective in reducing the risk of injury and illness and in improving conditions on construction projects. Work of this nature, coupled with advancements in engineering controls and other technological interventions, offers great potential for helping contractors manage safe projects and for protecting workers. This research should also focus on identifying the barriers to implementing effective safety culture and management systems. Research might incorporate behavioral psychology approaches and their application to culture change. The influx of Hispanic workers into the construction industry in the United States adds another level of complexity to this area of research.

Within the industry, many misconceptions exist about the higher cost of doing work more safely. An economic analysis of safety could focus on how products and processes within the marketplace can be rearranged to improve worker safety as well as contributing to an understanding of the downstream impacts of an enhanced safety culture. Research could also be undertaken to consider the role of the owner of a construction project in promoting a culture of safety and promoting good health practices to reduce the risks of injury and illness.

Increase Understanding of Construction Industry Organizational Factors (NORA2 Goal 10.0)

Goal 10.0 relates to the organizational complexity at construction worksites, which may involve multiple contractors and subcontractors, evolving production techniques, all operating independently but in close proximity. The intent of this goal is to promote research and stakeholder activities that will increase the knowledge of how construction industry organization and industry structure can influence safety and health performance.

Within this goal area, the committee recommends that the program increase its research focus on small contractors and on residential construction. Small businesses, which account for the majority of construction firms and workers, are typically more difficult to reach than are larger firms, yet safety and health risks among small businesses are often greater. Although a substantial part of the burden of morbidity and mortality occurs in residential construction, relatively little research specific to residential construction has been conducted. Residential construction poses a number of unique challenges to health and safety, including geographically dispersed worksites, a highly mobile workforce, and a relatively short build cycle that involves rapid changes in conditions on worksites. The small size of many residential contractors and in some areas of the country the transient and nonorganized labor force have been barriers to educational and research efforts. The interventions and safety guidelines may need to take a less traditional approach than those for the commercial, industrial, and heavy and civil engineering sectors.

Within the construction industry, many misconceptions exist about the higher cost of doing work more safely. An economic analysis of safety could focus on how products and processes within the marketplace can be rearranged to improve work safety as well as to improve understanding of the downstream impacts of an enhanced safety culture.

Improve Training and Its Dissemination (NORA2 Goal 11.0)

One area of opportunity for greater impact as determined by the committee's exchange with the stakeholder community is that of training to bring about an increased culture of safety within the construction industry. More attention is needed on how best to change workers' and contractors' safety behavior by changing their attitudes about safety and their understanding of related issues. Some work has been done in this area, notably in hearing conservation. A 1998 study reported that construction workers believed that they had developed occupational-noise-induced hearing loss (Lusk et al., 1998). The study showed a need for significant improvement in the consistent use of hearing protection among construction workers and a need to design hearing-conservation programs for the construction industry.

For many safety issues, such as trenching or ladder safety, the barrier to prevention is not the absence of knowledge, but the need for more widespread application of well-known safe work practices. To reach the many and decentralized target audiences with educational materials, the program should study the use of novel settings for dissemination, such as backhoe rental firms for trench safety education.

Increase Understanding of Risks to Vulnerable Worker Groups (NORA2 Goal 12.0)

In evaluating the Construction Research Program's research related to vulnerable populations (young workers, women, and immigrant workers) in the construction industry, the committee determined that for the most part the outputs of the program have addressed high-priority areas. However, intervention-related, measurable outcomes have been very limited, which may be an indication that program investigators, internal and external, have been unable to access adequate study populations, especially among the Hispanic workforce. In addition, there have not been reported evaluations of the literacy level, user friendliness, clarity, or design of the materials produced specifically for Hispanic workers.

The committee believes that the program should conduct and validate health and safety research on Hispanic workers that incorporates social, cultural, and external factors into the research design and implementation processes. Research on, dissemination to, and transfer activities for Hispanic workers will not be effective if informational materials are created by simply translating existing materials from English to Spanish. A clear plan for transfer activities and dissemination of research findings is necessary. In some cases it may be more effective to train Hispanic workers to teach safety-related training for this industry subpopulation than to use professional trainers who may not be attuned to social and cultural factors.

Role of Design (NORA2 Goal 13.0)

Improving the role of design in promoting health and safety is an important goal. Such research should focus on improving health and safety during all phases of a project, starting with the architect or engineer, continuing with the general contractor who maps out processes, and reaching the contractor who builds the project. Project design, construction processes, building materials, behaviors and work activities, tools, and fasteners could all be improved to promote greater health and safety. Three examples of such design interventions would be (1) designing residences so that fall-protection tie-off points are available during the construction process, (2) building gables and other roof structures on the ground and hoisting them into place rather than building them at height, and (3) using smaller and/or lighter blocks and sheets of building materials.

The Construction Research Program could also look at process issues, such as ways to reduce rework (and the potential for injuries and illnesses) through more focused work planning. Opportunities also exist for design improvement in work flow and methods to reduce worker exposures to hazardous materials, such as silica or welding fumes, by minimizing the number of times that a task is performed. Similarly, research could look at reorganizing tasks that require, for example, working overhead or in awkward positions, to reduce their frequency and duration.

An economic analysis of safety could focus on how products and processes within the marketplace can be rearranged to improve work safety and could also advance the understanding of the downstream impacts of an enhanced safety culture. The chemical industry's Responsible Care program is an example of this.

Improve Surveillance (NORA2 Goal 14.0)

The intent of Goal 14.0 is to improve surveillance in order to identify existing and emerging hazards and research needs and to evaluate intervention and organizational program effectiveness. One aspect of this goal is the development of "leading indicators" as opposed to "lagging indicators." An example of a leading indicator is information on the exposure or the existence of programs and practices that correlate with safety and health performance.

The NIOSH evidence package lists the various data series collected with the Safety Database and the Construction Industry and Construction Workforce Databases. These series seem to include most available data, although the series on Construction Market Trends from F.W. Dodge is a limited extraction of the available data. The raw data from the U.S. Department of Labor's Bureau of Apprenticeship Training (BAT) also appear to be missing. These training data would be useful to collect, as well as supplemental apprenticeship data from states such as California that do not feed their data into the BAT. Other data that might be useful include state-based data on the number of licensed craftspersons, for example, electricians or plumbers, once the data have been collected for several years. Government surveys on single-person firms and firms providing health insurance are also available. In both instances, either these data include construction as an industry category or it is possible to disaggregate the data into construction as an industry category. In short, more data on training, health insurance, licensing, and other factors might be collected.

Such data are important because training and experience are key factors in determining health and safety outcomes. Licensing and apprenticeship training are measures of training, and benefits, particularly health insurance, play a role in retaining workers in construction, which should influence health and safety outcomes.

The larger problem is that data designed to describe and help formulate better understanding of the workings of construction as they relate to health and safety outcomes require an analytical framing—that is, identification of the relevant data regarding construction that influence health and safety outcomes. For instance, it may be that the size of contractor firms influences safety outcomes or changes in safety outcomes. This question could be traced using available data from the Bureau of Labor Statistics (BLS) on the total injury rate by size of firms over time.

More generally, one of the main products of the collaboration between the Construction Research Program and the National Construction Center (NCC) in surveillance research is the *Construction Chart Book* (CPWR, 2007). This unique and helpful reference could be made stronger and more effective in assisting in surveillance efforts if it were more strongly rooted in an analytical framework based on the key factors influencing health and safety outcomes. This would entail framing safety within an economic, sociological, and organizational behavior analysis that would inform both what data were to be collected and how those data are presented.

The Construction Research Program staff and the NCC have identified two key weaknesses in national injury and illness statistics—namely, that illnesses in construction are not well canvassed by the BLS survey and that injuries are increasingly underreported. High turnover among firms and labor may account for the first weakness, while strategies for the avoidance of premiums for worker compensation abetted by the increased employment of undocumented workers in construction may account for the second weakness. Although neither the program nor the NCC is well positioned to remedy the need for raw data relative to these factors, more work could be done to estimate the degree of underreporting, particularly in the case of injuries, by relating reported injuries to changing worker compensation costs.

The program staff and the NCC are limited in their surveillance-based efforts to using data collected by others and in case studies. Accompanying these cost-based limitations are serious limitations in these second-party data. However, some of these limitations can be addressed. For instance, while there are no population-based surveys linking safety training with injury and illness outcomes, the increasing practice within some industries to prequalify contractors on the basis of their safety history or capabilities has the potential for providing these data on a limited-sample basis. Also, although there is no direct linkage between injury and illness data and workforce data, the fact that both types of data are gathered on a state-by-state basis provides the possibility for some linking of the two types. Given budgetary constraints and the constraints imposed by others on how data are collected, continued creativity in developing second-best approaches to making data cohere into a larger, better-linked picture of safety and construction should be encouraged.

Other suggestions for improvement include wrapping the gathering and presentation of surveillance data more thoroughly in a social science understanding of the causes of injuries and illnesses in construction in order to complement the engineering and epidemiological understanding of these events.

The program's work in demonstrating that elevated blood lead levels among construction workers can be reduced through a focused surveillance program also identifies an emerging research area. A substantial amount of work by the program, coupled with its influence on surveillance and worker protection regulations, offers an opportunity not only for the assessment of that influence but also an opportunity to improve worker safety and health. The focus of future work on improving color lead-detection wipes in the field will also provide an important and useful advancement in methods to ensure that potential lead exposures are identified. The program should also examine any differences in practices that may exist with work involving lead in commercial renovation and remodeling, and in steel structures other than bridges, and suggest any additional measures that may be necessary to protect construction workers.

Additional Areas for Future Research

In reviewing the Construction Sector Council goals, the NORA2 draft goals, and the NIOSH evidence package presented to the committee, the committee noted that a major gap across all areas is the paucity of well-performed intervention studies. The Construction Research Program has done excellent work in demonstrating the importance of risk factors and in suggesting interventions. The next step is to test these interventions through studies that measure the effectiveness of these interventions in construction work settings. Evaluating intervention effectiveness would help to validate the importance of risk factors and give more impetus to the dissemination of results by demonstrating the benefits of interventions. Evaluating the cost-effectiveness of specific safety and health measures would be particularly effective in speeding the dissemination of research to practice by demonstrating benefits to owners, contractors, and workers. That said, the committee recognizes that intervention studies are expensive and can be difficult to perform on construction worksites.

No additional future research appears to be contemplated by the program with respect to reducing exposures to asphalt in roofing operations, despite some success in identifying work practices that can reduce worker exposure to asphalt fumes. The committee recommends that the Construction Research Program examine the potential for building on its previous work to develop and implement engineering control approaches for roofing operations.

One other area not included in the program's future research activities is that of skin disease. In response to a modest research effort on this topic and its rather limited focus on Portland cement, the program has identified only a limited number of emerging issues targeted for continuing research in this area. These include developing a guide to working safely with epoxy resins (patterned after the work with Portland cement) and addressing the issue of potential ultraviolet exposures from welding operations. Both of these topics seem relevant for additional research. Overall, the committee is concerned that research on skin disorders is receiving insufficient attention. Exposures to chemicals that pose a hazard to the skin appear to be common across the construction industry. Skin disease appears to be an appropriate topic for inclusion in continuing research focusing on improving surveillance of hazards and outcomes.

OVERARCHING RECOMMENDATIONS FOR PROGRAM IMPROVEMENT

The overarching recommendations listed below apply to the Construction Research Program as a whole.

Transferring Research to Practice

Recommendation 1: Research-to-practice (R2P) efforts should involve individuals with the training or with the experience and skills to create strategic diffusion and social marketing plans for National Institute for Occupational Safety and Health research and to evaluate such plans' effectiveness.

Recommendation 2: Consideration should be given to having the majority of research-to-practice efforts of the Construction Research Program conducted through the National Construction Center.

A number of barriers currently exist within the program structure that limit the R2P efforts and likely their effectiveness. First, although the most recent cooperative agreement for the National Construction Center includes language stipulating that 20 percent of direct costs be used to increase the knowledge base for the effective diffusion of R2P for construction, in most cases this is not enough to implement more active dissemination strategies and evaluate their effectiveness. Outputs generated by external grantees and partners need to be included in the program's R2P efforts as well. Internally, program researchers have been encouraged to translate research findings to lay publications for target audiences and stakeholders. Indeed, an R2P plan is now a requirement for all internally funded projects.

The individual expertise called for by the committee in Recommendation 1 above does not necessarily need to reside in NIOSH staff. It could also be expertise within other government agencies such as the Centers for Disease Control and Prevention or OSHA or within the private sector, called on by the program to accomplish its diffusion goals more effectively.

Given that NIOSH is a federal agency, the document review process can be lengthy, and limitations are sometimes placed on what can and cannot be said, given that recommendations may be interpreted as policy. The National Construction Center is not constrained by these barriers, however, and thus consideration should be given to having the majority of R2P efforts conducted through the NCC; see Recommendation 2 above. This would allow the program staff to partner with NCC researchers and stakeholders and to focus on conducting the diffusion-related research necessary to determine the optimum ways to reach target audiences.

Resources

Recommendation 3: High-level attention should be given to determining how to provide program resources that are commensurate with a more robust pursuit of the Construction Research Program's goals.

Recommendation 4: The Construction Program Coordinator and the Construction Program Manager should both be devoted full-time to the Construction Research Program.

Recommendation 5: The National Construction Center should continue to be used as an important component in the Construction Research Program.

During its review, the committee concluded that in spite of budget constraints, the Construction Research Program has made an impact on one of the most dangerous and largest of U.S. industries. The total budget for the program from FY 1997 through FY 2007 has, in fact, stayed even or slightly declined in real dollars. It has also been declining as a portion of the total NIOSH budget during all of the period reviewed, 1996-2005. The committee finds the funding level inadequate and recommends that high-level attention be given to determining how to provide program resources that are commensurate with a more robust pursuit of the program's goals (see Recommendation 3, above).

A related matter, addressed in Recommendation 4 above, is that until very recently, NIOSH senior management had not made the commitment to assign at least one full-time senior-level staff person to coordinate the array of projects and activities carried out within the program. The committee supports NIOSH's action in making this a full-time position and recommends that NIOSH continue this practice into the future.

Until 2005, program activities were directed through the Construction Steering Committee, which is composed of representatives from NIOSH divisions and laboratories. In 2005, NIOSH appointed a senior lead team representative as the Construction Program Manager. The committee supports this action and recommends that this position also be devoted full-time to the Construction Research Program. The committee encourages NIOSH to ensure that this position has some level of budgetary authority and management responsibility so that the Construction Program Manager can provide strategic and programmatic leadership and also assist in holding the program accountable for achieving its future research goals.

As indicated in Recommendation 5 above, the committee also recommends that the National Construction Center continue to be used as an important component in NIOSH's Construction Research Program.

Increased Communication with Rule-Making Authorities

Recommendation 6: The Construction Research Program should establish a closer connection with the Occupational Safety and Health Administration and other regulatory or consensus standards organizations to help ensure that the program's research is applied effectively in rule-making efforts.

The committee recommends that the program increase its current level of communication with OSHA and other regulatory or consensus standards organizations about the evidence generated from its research activities. In addition to discussing research findings, program staff should communicate more fully on the economics of occupational disorders and illnesses and their impact on workers and contractors in the industry. Such information will provide valuable supporting documentation for recommendations made by the program with respect to regulatory action. Any role that the Construction Research Program can play in developing or strengthening standards that address risk exposure will likely increase its impact on risk reduction for occupational disorders and illnesses.

REFERENCES

CPWR (Center to Protect Workers' Rights). 2007. The Construction Chart Book: The U.S. Construction Industry and Its Workers. 4th ed. Washington, D.C.: Center for Construction Research and Training.

Lusk, S.J., M.J. Kauffman, and A. Sirkka. 1998. Use of hearing protection and perceptions of noise exposure and hearing loss among construction workers. American Industrial Hygiene Association Journal 59(7):466-470.

NIOSH (National Institute for Occupational Safety and Health). 2007. NIOSH Construction Research Program Evidence Package. Washington, D.C., July.

NORA (National Occupational Research Agenda) Construction Sector Council. 2008. National Occupational Research Agenda (NORA) Draft Preliminary Comment Version. Available at http://www.cdc.gov/niosh/nora/comment/public/constdraftdec2007/pdfs/constdraftdec2007.pdf. Accessed August 15, 2008.

Appendixes

A

Framework for the Review of Research Programs of the National Institute for Occupational Safety and Health*

This is the second version of a document prepared by the National Academies Committee for the Review of NIOSH Research Programs[1] also referred to as the Framework Committee. This document is not a formal report of the National Academies—rather, it is a framework proposed for use by multiple National Academies evaluation committees to review up to 15 National Institute for Occupational Safety and Health (NIOSH) research programs. It is a working document subject to modification by the Framework Committee on the basis of responses received from evaluation-committee members, NIOSH, stakeholders, and the general public during the course of the assessments.

---

*Version of 8/10/07.

[1]Members of the committee at the time this version was produced were David Wegman, *Chair* (University of Massachusetts Lowell School of Health and Environment), William Bunn III (International Truck and Engine Corporation), Carlos Camargo (Harvard Medical School), Susan Cozzens (Georgia Institute of Technology), Letitia Davis (Massachusetts Department of Public Health), James Dearing (Kaiser Permanente-Colorado), Fred Mettler, Jr. (University of New Mexico School of Medicine), Franklin Mirer (Hunter College School of Health Sciences), Jacqueline Nowell (United Food and Commercial Workers International Union), Raja Ramani (Pennsylvania State University), Jorma Rantanen (International Commission on Occupational Health), Rosemary Sokas (University of Illinois at Chicago School of Public Health), Richard Tucker (Tucker and Tucker Consultants, Inc., and University of Texas at Austin), and James Zuiches (North Carolina State University). Sammantha Magsino (National Academies staff) was the study director. Joseph Wholey (University of Southern California), former committee member, contributed to the first version of this document. Part V includes brief biographies of current committee members.

This version reflects several significant changes to the original framework document (version 12/19/05) that was used to guide the work of the first four evaluation committees (Hearing Loss; Mining; Agriculture, Forestry, and Fishing; and Respiratory Disease). Changes were made in response to feedback from members and staff of these committees, as well as other comments on the original framework, in order to make the document more useful to evaluation committees as they carry out their work. In particular, the following changes were made to the framework document during the revision process:

- The wording of some of the relevance and impact scores were edited to make the wording more precise and to reduce situations where the original scores were non-unique or overlapping (revised scoring criteria are given in Boxes 2 and 3).
- A new table was added to provide explicit guidance to evaluation committees on how to weigh differences in the observed levels of "research priority" and "engagement in appropriate transfer activities" in arriving at a single integer score for relevance (see Table 6).
- The guidance on scoring was clarified to make more explicit that all scores are to be given as integers.
- The NIOSH logic model was updated (see Figure 1).
- The table on evaluation committee information needs (Table 2) was reorganized to be more consistent with the NIOSH logic model, and additional information needs identified by the first set of evaluation committees were added.
- A worksheet to assist with the development of scores has been deleted and key components of the worksheet have been incorporated into appropriate sections throughout the document.
- The organization of the document was modified to more closely follow the revised statement of task and to improve readability.
- A number of sections of text originally presented in outline form were modified in tables or boxes to make the information more accessible.

This second version of the framework document remains a working document subject to further modification by the Framework Committee on the basis of input received from evaluation committee members, NIOSH, stakeholders, and the general public during the course of the assessments.

CONTENTS

ABBREVIATIONS AND ACRONYMS

ABLES	Adult Blood Lead Epidemiology and Surveillance
AOEC	Association of Occupational and Environmental Clinics
BLS	Bureau of Labor Statistics
CDC	Centers for Disease Control and Prevention
CSTE	Council of State and Territorial Epidemiologists
DOD	U.S. Department of Defense
EC	Evaluation Committee
EPA	Environmental Protection Agency
FACE	Fatality Assessment Control and Evaluation
FC	Framework Committee
HHE	Health Hazard Evaluation
MSHA	Mine Safety and Health Administration
NIH	National Institutes of Health
NIOSH	National Institute for Occupational Safety and Health
NORA	National Occupational Research Agenda
NORA1	National Occupational Research Agenda 1996-2005
NORA2	National Occupational Research Agenda 2005-forward
OSH Review Commission	Occupational Safety and Health Review Commission
OSHA	Occupational Safety and Health Administration
OSHAct	Occupational Safety and Health Act of 1970
PART	Performance Assessment Rating Tool
PEL	permissible exposure limit
RFA	request for applications
SENSOR	Sentinel Event Notification System of Occupational Risks

TMT	tools, methods, or technologies
USDA	U.S. Department of Agriculture

I. INTRODUCTION

In September 2004, the National Institute for Occupational Safety and Health (NIOSH) contracted with the National Academies to conduct a review of NIOSH research programs. The goal of this multiphase effort is to assist NIOSH in increasing the impact of its research efforts that are aimed at reducing workplace illnesses and injuries and improving occupational safety and health. The National Academies assigned the task to the Division on Earth and Life Studies and the Institute of Medicine.

The National Academies appointed a committee of 14 members, including persons with expertise in occupational medicine and health, industrial health and safety, industrial hygiene, epidemiology, civil and mining engineering, sociology, program evaluation, communication, and toxicology; representatives of industry and of the workforce; and a scientist experienced in international occupational-health issues. The Committee on the Review of NIOSH Research Programs, referred to as the Framework Committee (FC), prepared the first version of this document during meetings held on May 5-6, July 7-8, and August 15-16, 2005. This second version was finalized after the Framework Committee's May 30-31, 2007, meeting, based on feedback received on the framework from the first two independent evaluation committees, NIOSH leadership, and National Academies' staff, as well as discussions during an earlier FC meeting in April 2006.

This document is not a report of the National Academies; rather, it presents the evaluation framework developed by the FC to guide and provide common structure for the reviews of as many as 15 NIOSH programs during a 5-year period by independent evaluation committees (ECs) appointed by various divisions and boards of the National Academies. It is a working document to be shared with NIOSH and the public. This version has been modified by the FC on the basis of responses from the ECs, NIOSH, NIOSH stakeholders, and the public; and it may be modified again. It is incumbent on the ECs to consult with the FC if portions of the evaluation framework presented here are inappropriate for specific programs under review.

I.A. Overview of Charge to Evaluation Committees

At the first meeting of the FC, Lewis Wade, NIOSH senior science adviser, emphasized that a review of a NIOSH program should focus on the program's relevance to and impact on health and safety in the workplace. In developing a framework, the FC considered the following elements of the charge to the ECs:

1. Assessment of the program's contribution, through occupational safety and health research, to reductions in workplace hazardous exposures, illnesses, or injuries through

    a. an assessment of the relevance of the program's activities to the improvement of occupational safety and health, and
    b. an evaluation of the impact that the program's research has had in reducing work-related hazardous exposures, illnesses, and injuries.

    The evaluation committee will rate the performance of the program for its relevance and impact using an integer score of 1-5. Impact may be assessed directly (for example, on the basis of reductions in illnesses or injuries) or, as necessary, by using intermediate outcomes to estimate impact. Qualitative narrative evaluations should be included to explain the numerical ratings.
2. Assessment of the program's effectiveness in targeting new research areas and identifying emerging issues in occupational safety and health most relevant to future improvements in workplace protection. The committee will provide a qualitative narrative assessment of the program's efforts and suggestions about emerging issues that the program should be prepared to address.

I.B. Evaluation Committees

Individual ECs will be formed in accordance with the rules of the National Academies for the formation of balanced committees. Each EC will comprise persons with expertise appropriate for the specific NIOSH research program under review and may include representatives of stakeholder groups (such as labor unions and industry), experts in technology and knowledge transfer, and program evaluation. The EC will gather appropriate information from the sponsor (the NIOSH research program under review), stakeholders affected directly by NIOSH program research, and relevant independent parties. Each EC will consist of about 10 members, will meet about three times, and will prepare a report. The National Academies will deliver the report to NIOSH within 9 months of the first meeting of the EC. EC reports are subject to the National Academies report-review process.

I.C. NIOSH Strategic Goals and Operational Plan

As a prelude to understanding the NIOSH strategic goals and operational plan, NIOSH research efforts should be understood in the context of the Occupational

Safety and Health Act (OSHAct), under which it was created. The OSHAct identifies workplace safety and health as having high national priority and gives employers the responsibility for controlling hazards and preventing workplace injury and illness. The act creates an organizational framework for doing that, assigning complementary roles and responsibilities to employers and employees, the Occupational Safety and Health Administration (OSHA), the states, the Occupational Safety and Health (OSH) Review Commission, and NIOSH. The act recognizes NIOSH's role and responsibilities to be supportive and indirect. NIOSH research, training programs, criteria, and recommendations are intended to be used to inform and assist those more directly responsible for hazard control (OSHAct Sections 2b, 20, and 22).

Section 2b of the OSHAct describes 13 interdependent means of accomplishing the national goal, one of which is "by providing for research . . . and by developing innovative methods . . . for dealing with occupational safety and health problems." Sections 20 and 22 give the responsibility for that research to NIOSH. NIOSH is also given related responsibilities, including the development of criteria to guide prevention of work-related injury or illness; development of regulations for reporting on employee exposures to harmful agents; establishment of medical examinations, programs, or tests to determine illness incidence and susceptibility; publication of a list of all known toxic substances; assessment of potential toxic effects or risks associated with workplace exposure in specific settings; and conduct of education programs for relevant professionals to carry out the OSHAct purposes. NIOSH is also responsible for assisting the secretary of labor regarding education programs for employees and employers in hazard recognition and control.

The NIOSH mission is "to provide national and world leadership to prevent work-related illness, injury, disability, and death by gathering information, conducting scientific research, and translating the knowledge gained into products and services." To fulfill its mission, NIOSH has established the following strategic goals:[2]

- **Goal 1: Conduct research to reduce work-related illnesses and injuries.**
    - Track work-related hazards, exposures, illnesses, and injuries for prevention.
    - Generate new knowledge through intramural and extramural research programs.

[2]See http://www.cdc.gov/niosh/docs/strategic/.

  - Develop innovative solutions for difficult-to-solve problems in high-risk industrial sectors.

- **Goal 2: Promote safe and healthy workplaces through interventions, recommendations, and capacity building.**
  - Enhance the relevance and utility of recommendations and guidance.
  - Transfer research findings, technologies, and information into practice.
  - Build capacity to address traditional and emerging hazards.

- **Goal 3: Enhance global workplace safety and health through international collaborations.**
  - Take a leadership role in developing a global network of occupational health centers.
  - Investigate alternative approaches to workplace illness and injury reduction and provide technical assistance to put solutions in place.
  - Build global professional capacity to address workplace hazards through training, information sharing, and research experience.

In 1994, NIOSH embarked on a national partnership effort to identify research priorities to guide occupational health and safety research for the next decade. The National Occupational Research Agenda (NORA) identified 21 high-priority research subjects (see Table 1). The NORA was intended not only for NIOSH but for the entire occupational health community. In the second decade of the NORA, NIOSH is working with its partners to update the research agenda, using an approach based on industry sectors. NIOSH and its partners are working through sector research councils to establish sector-specific research goals and objectives. The emphasis is on moving research to practice in workplaces through sector-based partnerships.

Figure 1 is the NIOSH operational plan, presented as a logic model,[3] of the path from inputs to outcomes for each NIOSH research program. The FC adapted the model to develop its framework. NIOSH will provide similar logic models appropriate to each research program evaluated by an EC.

[3]Developed by NIOSH with the assistance of the RAND Corporation.

TABLE 1 NORA High-Priority Research by Category

Category	Priority Research Area
Disease and injury	Allergic and irritant dermatitis
	Asthma and chronic obstructive pulmonary disease
	Fertility and pregnancy abnormalities
	Hearing loss
	Infectious diseases
	Low-back disorders
	Musculoskeletal disorders of upper extremities
	Trauma
Work environment and workforce	Emerging technologies
	Indoor environment
	Mixed exposures
	Organization of work
	Special populations at risk
Research tools and approaches	Cancer research methods
	Control technology and personal protective equipment
	Exposure-assessment methods
	Health-services research
	Intervention-effectiveness research
	Risk-assessment methods
	Social and economic consequences of workplace illness and injury
	Surveillance research methods

I.D. Evaluation Committees' Information Needs

Each NIOSH program under review will provide information to the relevant EC, including that outlined in Table 2. The EC may request additional information of NIOSH as needed, and NIOSH should provide it as quickly as is practical. NIOSH should consider organizing the information listed in Table 2 by subprogram or program as appropriate and to the extent possible.

In addition to the information provided by NIOSH, the EC should independently collect additional information that it deems necessary for evaluation (for example, the perspectives of stakeholders, such as OSHA, MSHA, unions and workforces, and industry). In conducting the review, the EC should continually examine how individual projects or activities contribute to the impact and relevance of a program as a whole.

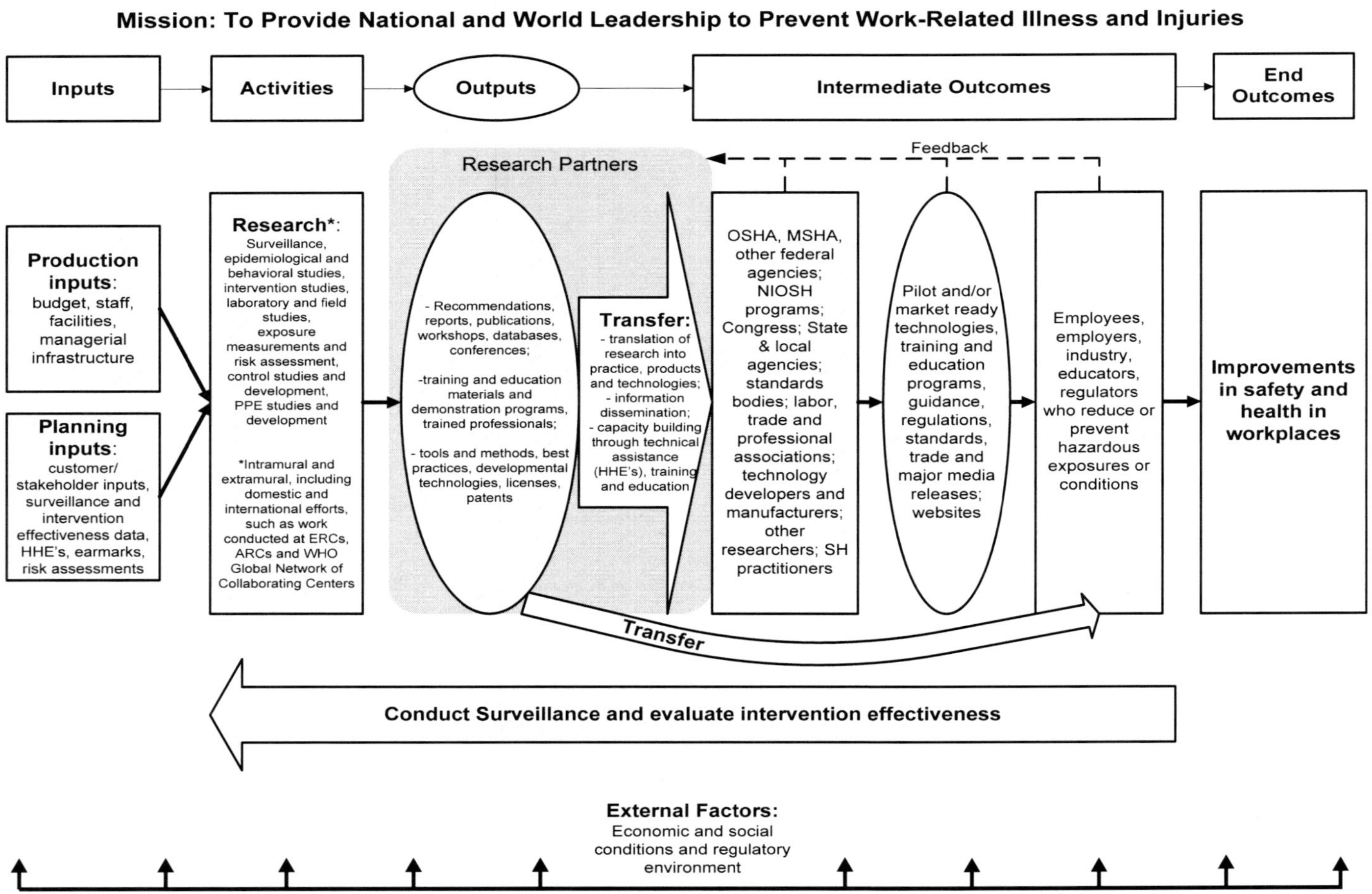

FIGURE 1 The NIOSH operational plan presented as a logic model.

TABLE 2 Evaluation Committee Information Needs

- **Program background and resources:**
  - Program history.
  - Major program challenges.
  - Program strategic goals and objectives, past (for period under review) and current.
  - Major subprograms (if appropriate).
  - Results of previous program reviews (for example, annual review by NIOSH leadership team or external scientific program reviews).
  - External factors affecting the program.

- **Interactions with stakeholders and with other NIOSH programs:**
  - The role of program research staff in NIOSH policy-setting, OSHA and MSHA standard-setting, voluntary standard-setting and other government policy functions.
  - Interactions and working relationships with other NIOSH programs.
  - Identification of other institutions and research programs with overlapping or similar portfolios and an explanation of the relationship between NIOSH activities and those of other institutions.
  - Key partnerships with employers, labor, other government organizations, academic institutions, nonprofit organizations, and international organizations.

- **Program inputs:**
  - Program resources (also called *production inputs*).
    - Funding by year for period under review.
    - Funding by objective or subprogram.
    - Program staffing, FTE's, and laboratory facilities, by subprogram (if indicated).
    - Percentage of program budget that is discretionary (beyond salaries).
    - Percentage of program budget that is earmarked.
    - Contributions from other agencies (in kind or funds).
  - Planning inputs.
    - Surveillance data, inputs from the Health Hazard Evaluation (HHE) or Fatality Assessment Control and Evaluation (FACE) program, or intramural and extramural research findings that influenced program goals and objectives.
    - Planning inputs from stakeholders, for example, advisory groups, NORA teams, and professional, industry, and labor groups (specify if any input from groups representing small business or vulnerable populations).
    - Related OSHA, Mine Safety and Health Administration (MSHA) strategic plans, or other input.
    - Process for soliciting and approving intramural research ideas.
    - Process for soliciting and approving program-supported extramural research activities.

- **Program activities (more details provided in Table 3):**
  - Intramural.
    - Surveillance activities.
    - Research activities (projects).

continued

TABLE 2 Continued

    - Transfer activities to encourage implementation of research results for improved occupational safety and health (for example, information dissemination, technical assistance, and technology and knowledge transfer).
    - Key collaborations in intramural activities (for example, with other government agencies, academe, industry, and unions).
  - Extramural funded by NIOSH.
    - Requests for applications (RFAs) developed by program.
    - Funded projects: grants, cooperative agreements, and contracts, such as
      - ◊ Surveillance activities.
      - ◊ Research activities.
      - ◊ Transfer activities.
      - ◊ Capacity-building activities.

- **Outputs (products of the research program—more details provided in Table 4):**
  - Intramural.
    - Peer-reviewed publications, agency reports, alerts, and recommendations.
    - Databases, Web sites, tools, and methods (including education and training materials).
    - Technologies developed and patents.
    - Sponsored conferences and workshops.
  - Extramural (to the extent practical).

- **Intermediate outcomes:**
  - Standards or guidelines issued by other agencies or organizations based in whole or in part on NIOSH research.
  - Adoption and use of control or personal protective technologies developed by NIOSH.
  - Evidence of industry, employer, or worker behavioral changes in response to research outputs.
  - Use of NIOSH products by workers, industry, occupational health and safety professionals, health care providers, and so on (including internationally).
  - NIOSH Web-site hits and document requests.
  - Unique staff or laboratory capabilities that serve as a national resource.
  - Other intermediate outcomes.

- **End outcomes:**
  - Data on program impact on rates and numbers of injuries and illnesses and exposures in the workplace (including trend data, if available).
  - Documentation of workplace risk reduction (quantitative, qualitative, or both).

- **Description of current processes for setting research priorities and identifying emerging issues in the workplace.**

I.E. Prior Evaluations

Several NIOSH programs have already been evaluated by internal and external bodies. The evaluations may have been part of an overall assessment of NIOSH, such as the 2005 Performance Assessment Rating Tool (PART) review,[4] or the evaluation of specific research program elements, such as any external scientific-program review. NIOSH should inform of, and the ECs should review, all prior evaluations of the program under review as an aid to understanding the evolution of the program and its elements. The EC evaluations, however, are independent of prior reviews and evaluations.

II. SUMMARY OF EVALUATION PROCESS

The ECs will assess the relevance and impact of NIOSH research programs. In conducting their evaluations, the ECs should ascertain whether NIOSH is doing the right things (relevance) and whether these things are improving health and safety in the workplace (impact).

II.A. The Evaluation Flowchart (Figure 2)

To address its charge, the FC simplified the logic model of Figure 1 into a flowchart (Figure 2) that breaks the NIOSH logic model into discrete, sequential program components to be assessed by the EC. Each component of Figure 2 is addressed in greater detail in the indicated section of this document. The FC understands that the activities of any research program will not be as linear as presented in either Figures 1 or 2. The major components to be evaluated are

- major program *challenges,*
- strategic *goals and objectives,*
- *inputs* (such as budget, staff, facilities, the institute's research management, the NIOSH Board of Scientific Counselors, the NORA process, and NORA work groups),
- *activities* (efforts by NIOSH staff, contractors, and grantees, such as hazard surveillance; surveillance for injury, illness, and biomarkers of effect; exposure-measurement research; safety-systems research; injury-prevention research; health-effects research; intervention

[4]The PART focuses on assessing program-level performance and is one of the measures of success of the budget and performance integration initiative of the president's management agenda (see CDC Occupational Safety and Health at http://www.whitehouse.gov/omb/budget/fy2006/pma/hhs.pdf).

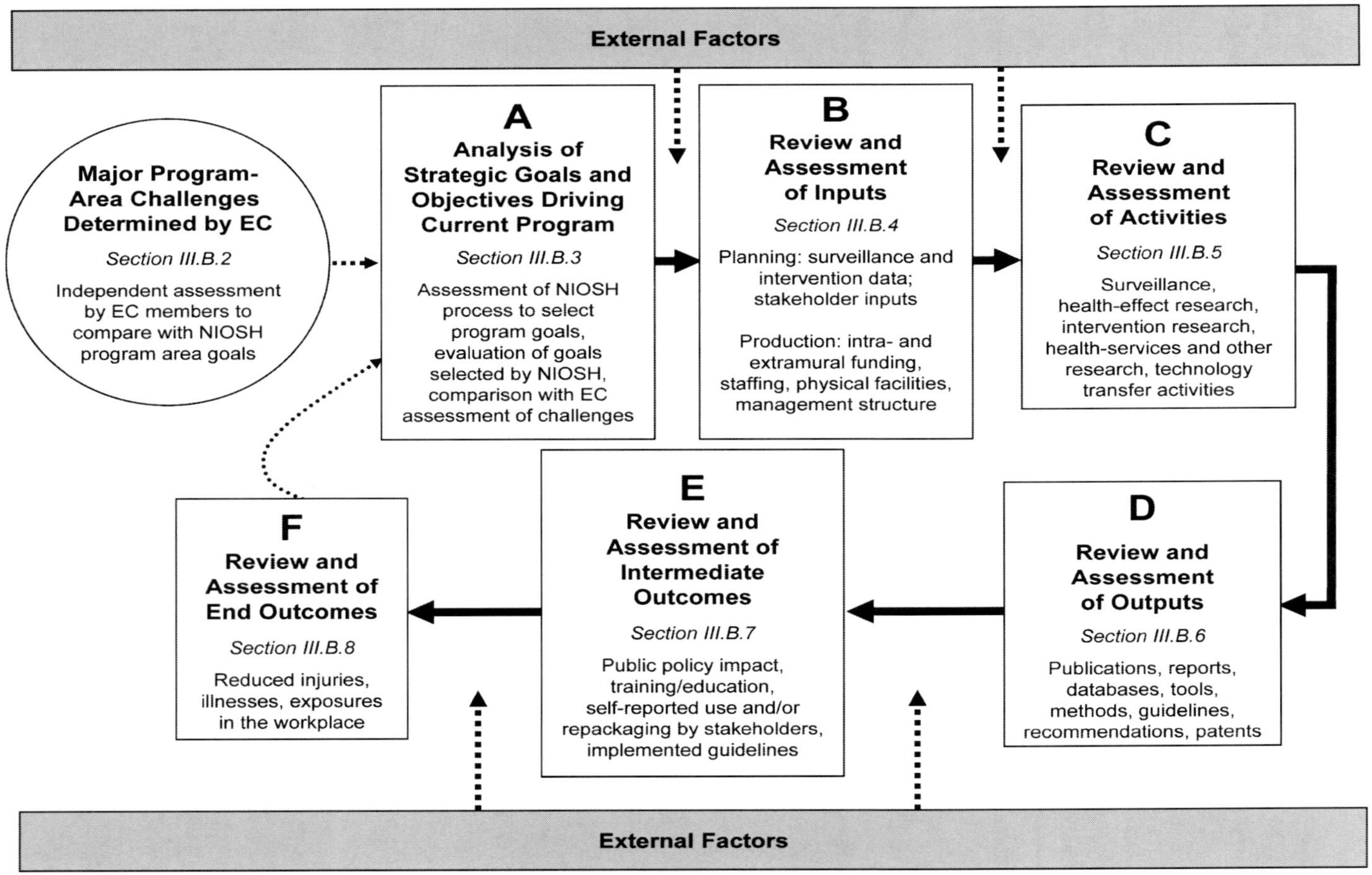

FIGURE 2 Flowchart for the evaluation of the NIOSH research program.

research; health-services research; and technology and knowledge transfer activities),

- *outputs* (NIOSH products, such as publications, reports, conferences, databases, tools, methods, guidelines, recommendations, education and training, and patents),
- *intermediate outcomes* (responses by NIOSH stakeholders to NIOSH products, such as public or private policy change, training and education in the form of workshop or seminar attendance, self-reported use or repackaging of NIOSH data by stakeholders, adoption of NIOSH-developed technologies, implemented guidelines, licenses, and reduction in workplace hazardous exposure), and
- *end outcomes* (such as reduction in work-related injuries or illnesses or hazardous exposures in the workplace).

The flowchart summarizes the FC's vision of how a program evaluation should occur. In evaluating each program or major subprogram, the EC must collect, analyze, and evaluate information on items described in each of the boxes of Figure 2, regardless of management structure (such as linear or matrix). The FC recognizes that the components of any program will not fit perfectly in any category in Figure 1 or 2. For example, training and development programs were appropriately defined as outputs by NIOSH in the logic model (Figure 1), but the FC finds more value in focusing on the responses to these outputs as intermediate outcomes (Figure 2, Box E) in the flowchart. The committee further recognizes that matrix organizations may have little control over the input portion of the logic model and that matrix program management may have fewer resources of its own on which to base its decisions. Following the suggested evaluation procedures, however, should ensure a desired level of consistency and comparability among all the ECs.

Drawing on the program logic model, the flowchart, and EC members' expertise, the ECs will delineate important inputs and external factors affecting the NIOSH research program's agenda and the consequences of NIOSH research activities. Examples of external factors are research activities of industry and other federal agencies and the political and regulatory environment. For purposes of this review, the results of inputs and external factors are the program research activities, outputs, and associated transfer activities that may result in intermediate outcomes and possibly end outcomes.

II.B. Steps in Program Evaluation

The FC concludes that useful evaluation requires a disciplined focus on a small number of questions or hypotheses typically related to program goals, performance criteria, and performance standards; a rigorous method of answering the questions or testing the hypotheses; and a credible procedure for developing qualitative and quantitative assessments. The evaluation process developed by the FC is summarized in Box 1 and described in detail in Section III of this document.

III. EVALUATION OF A NIOSH RESEARCH PROGRAM—THE PROCESS

III.A. Analysis of External Factors Relevant to the NIOSH Research Program

As depicted in the logic model (Figure 1), reduction in injury and illness (end outcomes) or in exposure (intermediate outcome) is affected by stakeholder activities (external factors). Actions of those in labor, industry, regulatory entities, and others beyond NIOSH's control are necessary for the implementation of NIOSH recommendations. Implementation of research findings may depend on existing or future policy considerations.

External factors may be considered as forces beyond the control of NIOSH that may affect the evolution of a program. External factors influence NIOSH's progress through all phases of the logic model and flowchart; from inputs to end outcomes (see Figures 1 and 2). Identification of external factors by an EC is essential because it provides the context for evaluation of the NIOSH program. External factors may be best assessed on the basis of the expert judgment of EC members who have knowledge of the field of research. Information regarding external factors should also be sought from NIOSH, OSHA, and MSHA staff and from other stakeholders. The EC, however, may choose additional approaches to assess external factors. NIOSH should identify and describe external factors early in the evaluation sequence (see Table 2). Factors external to NIOSH might have been responsible for achieving some outcomes or might have presented formidable obstacles. The EC must address both possibilities.

Some external factors may involve constraints on research activities related to target populations, methodologic issues, and resource availability. ECs might examine whether or not the following are true:

- Projects addressing a critical health need are technologically feasible. However, a workforce of appropriate size and with appropriate duration and distribution of exposure for measuring a health effect may not exist; for example, no population of workers has been exposed for 30 years to formaldehyde at the current OSHA permissible exposure limit (PEL), so the related cancer mortality cannot yet be directly assessed.
- Research is inhibited because NIOSH investigators are unable to access an adequate study population. Under current policy, NIOSH must

BOX 1
The Evaluation Process

1. Gather appropriate information from NIOSH and other sources (see Table 2).

2. Determine timeframe to be covered in the evaluation (see III.B.1.).

3. Identify major program area challenges and objectives (see III.B.2.).
All NIOSH research programs are designed to be responsive to present or future workplace safety and health issues. Each research program should have its own objectives. Each EC will provide an independent assessment of the major workplace health and safety problems related to the program under review and determine whether they are consistent with the program's stated goals and objectives.

4. Identify subprograms and major projects in the research program.
Each EC must determine how to disaggregate a program to achieve a manageable and meaningful evaluation of its components, and of the overall program. A program may need to be broken down into several recognizable subprograms or major projects if an effective evaluation is to be organized. It may be advantageous for an EC to disaggregate a program into subprograms that NIOSH identifies.

5. Evaluate the subprogram components sequentially (see III.B.2. through III.B.8.), using the flowchart (Figure 2) as a guide.
This will involve a qualitative assessment of each component of the research program. ECs will use professional judgment to answer questions and follow the guidance provided by the FC.

6. Evaluate the research program's potential outcomes that are not yet appreciated (see III.B.9.).

7. Evaluate the important subprogram outcomes specifically for contributions to improvements in workplace safety and health.
Guidance is provided with specific items for consideration (see III.B.10.).

8. Evaluate and score the overall program for *relevance* (see III.B.10.).
Final program ratings will consist of an integer score and discussion of its rationale.

9. Evaluate and score the overall program for *impact* (see III.B.10.).
Final program ratings will consist of an integer score and discussion of its rationale.

10. Identify success in targeting priority research and emerging issues (see III.C.).
The EC should briefly discuss its assessment of the NIOSH program's process for determining priorities for research and emerging workplace issues. The ECs should also independently identify emerging workplace issues for which the NIOSH program under review should be prepared.

11. Prepare report by using the template provided in Section IV as a guide.

either obtain an invitation by management to study a workplace or seek a judicial order to provide authority to enter a worksite. (Cooperation under court order may well be insufficient for effective research.)

- Research is inhibited because the work environment, materials, and historical records cannot be accessed even with management and workforce cooperation.
- Adequate or established methods do not exist for assessing the environment.
- The NIOSH contribution to a particular field of research is reduced because other institutions are working in the same field.
- NIOSH resources are inadequate to tackle key questions.

Evaluation of the impact of NIOSH research outputs on worker health and safety may require consideration of external factors that might impede or aide implementation, measurement, and so on. ECs might consider whether or not the following are true:

- Regulatory end points are unachievable because of obstacles to regulation or because of differing priorities of the regulatory agencies. For example, there may be no implementation of recommendations for improved respiratory protection programs for health-care workers because of enforcement policies or lack of acceptance by the health-care institution administrators.
- A feasible control for a known risk factor or exposure is unimplemented because the costs of implementation are too high or because current economic incentives do not favor such actions.
- End outcomes are unobservable because baseline and continuing surveillance data are not available. For example, the current incidence of occupational noise-induced hearing loss is not known although surveillance for a substantial threshold shift is feasible. (NIOSH conducts surveillance of work-related illnesses, injuries, and hazards, but comprehensive surveillance is not possible with existing resources.)
- Reductions in adverse effects of chronic exposure cannot be measured. For example, 90% of identified work-related mortality is from diseases, such as cancer, that arise only after decades of latency after first exposure; therefore, effects of reducing exposure to a carcinogen cannot be observed in the timeframe of most interventions.
- A promulgated regulation requires a technology that was developed but not widely used.

- Reductions in fatal traumatic injuries occur because more-hazardous manufacturing jobs are replaced by less-hazardous knowledge-based jobs.

III.B. Evaluating NIOSH Research Programs by Using the Flowchart

The FC used the NIOSH logic model (Figure 1) to define the scope and stages of an EC evaluation. The evaluation of the elements in the flowchart (Figure 2) summarizes the FC's vision of how a program evaluation should proceed. FC members also identified numerous possible factors to consider in assessing the relevance of NIOSH research-program components, including the following:

- The severity or frequency of health and safety hazards addressed and the number of people at risk (magnitude) for these hazards.
- The extent to which NIOSH research programs identify and address gender-related issues and issues of vulnerable populations: Vulnerable populations are defined as groups of workers who have biologic, social, or economic characteristics that place them at increased risk for work-related conditions or on whom inadequate data have been collected. Vulnerable populations include disadvantaged minorities, disabled persons, low-wage workers, and non-English-speakers for whom language or other barriers present health or safety risks.
- The extent to which NIOSH research programs address the health and safety needs of small businesses.
- The "life stage" of problems being addressed: As the health effects are understood, efforts should shift to intervention research, from efficacy to intervention, and to intervention-effectiveness research. Gaps in the spectrum of prevention need to be addressed; for example, research on exposure assessment may be necessary before the next intervention steps can be taken.
- The structure, in addition to the content, of the research program: A relevant research program is more than a set of unrelated research projects; it is an integrated program involving interrelated surveillance, research, and transfer activities.
- Appropriate NIOSH consideration of stakeholder input.

The ECs may consider those and other important factors that bear on relevance as they progress through each stage of an evaluation.

The following subsections are intended to guide the EC through the evaluation process and flowchart in Figure 2. Each begins with a definition of the component being evaluated, provides questions for the EC to consider during the course of its evaluation, and provides some guidance regarding the assessment of the component. The FC admittedly provides little guidance regarding the evaluation of programs that are organized in a matrix structure or programs that have large extramural research components. Because of the uniqueness of each program, each EC must determine the most reasonable way to apply the criteria established in this document.

III.B.1. Identifying the Period for Evaluation

By studying materials presented by the NIOSH research program and other sources, the EC will become familiar with the history of the research program being evaluated and its major subprograms, goals, objectives, resources, and other pertinent information. Having that information, the EC should choose the period most appropriate for the evaluation. EC efforts should focus on the impact and relevance of the NIOSH program in the most recent appropriate period. As a starting point, the ECs might consider three general timeframes:

- 1970-1995, the period from the founding of NIOSH to the initiation of NORA (pre-NORA period)
- 1996-2005 (NORA 1 period)
- After 2005 (NORA 2 period)

Those timeframes are provided as general guidance; the period chosen for review will take into consideration suggestions from the NIOSH research program under review. It is recognized that many of the intermediate and end outcomes documented since 1996 are consequences of research outputs completed before 1996.

III.B.2. Identifying Major Challenges (Figure 2, Circle)

Early in the assessment process, the EC itself should identify the major workplace health and safety challenges for the research program under review. In arriving at a list of challenges, the EC should rely on surveillance findings, including those of NIOSH investigations of sentinel events (through health-hazard or fatality-assessment programs), external advisory inputs, and its own expert judgment. The EC will then be able to compare its own assessment of workplace challenges with the NIOSH program goals and objectives. The congruence between the two will be useful during the assessment of relevance.

III.B.3. Analysis of Research-Program Strategic Goals and Objectives (Figure 2, Box A)

The research program goals and objectives should be evaluated with a focus on how each program goal is related to NIOSH's agency wide strategic goals and to the program challenges identified in the step above (Section III.B.2.). The importance or relevance of an issue may differ from the influence of NIOSH-funded research in addressing it. The EC should recognize that NIOSH research priorities may be circumstantial (for example, congressionally funded) rather than based on NIOSH's assessment of the state of knowledge.

Questions to Guide the Evaluation Committee

1. Are the strategic goals and objectives of the program well defined and clearly described?
2. How well were program goals and objectives aligned with NORA 1 priorities during the last decade?
3. How are current program strategic goals and objectives related to current NIOSH strategy, including NORA 2?
4. Are the research program goals, objectives, and strategies relevant to the major challenges for the research program and likely to address emerging problems in the research program (as determined by the EC while addressing Section III.B.2.)?
    a. Did past program goals and objectives (research and dissemination and transfer activities) focus on the most relevant problems and anticipate the emerging problems in the research program?
    b. Do the current program goals and objectives target the most relevant problems?

Assessment

The EC should provide a qualitative assessment that discusses the relevance of the program's goals, objectives, and strategies in relation to its major challenges.

III.B.4. Review of Inputs (Figure 2, Box B)

Planning inputs include input from stakeholders, surveillance and intervention data, and risk assessments. Production inputs include intramural and extramural funding, staffing, management structure, and physical facilities.

The EC should examine existing intramural and extramural resources and, potentially, prior surveys or case studies that might have been developed specifically to assess progress in reducing workplace illnesses and injuries and to provide information relevant to the targeting of research to future needs. The NIOSH research program should provide the EC all relevant planning and production inputs (see below and Table 2 for examples).

Planning Inputs

Planning inputs can be qualitative or quantitative. Sources of qualitative inputs include the following:

- Federal advisory committees (such as the Board of Scientific Counselors, the Mine Safety and Health Research Advisory Committee, and the National Advisory Committee on Occupational Safety and Health)
- NORA research partners, initial NORA stakeholder meetings, later NORA team efforts (especially strategic research plans), and the NORA Liaison Committee and federal liaison committee recommendations
- Industry, labor, academe, professional associations, industry associations, and the Council of State and Territorial Epidemiologists (CSTE)
- OSHA and MSHA strategic plans and other federal research agendas

Attention should be given to how comprehensive the inputs have been and to what extent gaps in input have been identified and considered by NIOSH.

Sources of quantitative inputs include the following:

- Intramural surveillance information, such as descriptive data on exposures and outcomes (appropriate data may be available from a number of NIOSH divisions and laboratories)
- HHEs
- Reports from the FACE program
- Extramural health-outcome and exposure-assessment data from OSHA, MSHA (both safety and health inspection data), the Bureau of Labor Statistics, the U.S. Department of Defense (DOD), and the U.S. Department of Agriculture (USDA) (fatality, injury, and illness surveillance data); state government partners, including NIOSH-funded state surveillance programs, such as Sentinel Event Notification System of Occupational Risks (SENSOR), Adult Blood Lead Epidemiology and

Surveillance (ABLES), and state-based FACE; and nongovernment organizations, such as the National Safety Council, the Association of Occupational and Environmental Clinics (AOEC), the American Society of Safety Engineers, and the American College of Occupational and Environmental Medicine
- Appropriate data from investigator-initiated extramural research funded by NIOSH

Production Inputs

For the research program under review, NIOSH should identify portions of the NIOSH intramural budget, staff, facilities, and management that play major roles in the research program. Production inputs should be described primarily in terms of intramural research projects, relevant extramural projects (particularly cooperative agreements and contracts), HHEs, and related staff. Consideration should also be given to leveraged funds provided by such partners as the National Institutes of Health (NIH) and the Environmental Protection Agency (EPA) for joint requests for applications or program announcements; and to OSHA, MSHA, and U.S. Department of Defense (DOD) contracts with NIOSH.

Assessment of inputs should include EC consideration of the degree to which allocation of funding and personnel was commensurate with the resources needed to conduct the research and the extent to which funding for the relevant intramural research activity has been limited by lack of discretionary spending beyond salaries (travel, supplies, external laboratory services, and so on). Thus, assessments should consider the adequacy of the qualitative and quantitative planning and production inputs, given the tasks at hand.

Questions to Guide the Evaluation Committee

1. Do planning, production, and other input data promote program goals?
2. How well are major planning, production, and other program inputs used to support the major activities?
3. Is input obtained from stakeholders, including input representing vulnerable working populations and small businesses?
4. Are production inputs (intramural and extramural funding, staffing, management, and physical infrastructure resources) consistent with program goals and objectives?

Assessment

The EC should provide a qualitative assessment that discusses the quality, adequacy, and use of inputs.

III.B.5. Review of Activities (Figure 2, Box C)

Activities are defined as the efforts and work of a program's staff, grantees, and contractors. For present purposes, activities of the NIOSH program under review are divided into research and transfer activities. Table 3 is intended to guide the EC and NIOSH as to the type and organization of information required to evaluate program activities. The table may be incomplete, and some types of research activity may not be applicable to a given NIOSH program. Research activities include safety research, health-outcomes research, safety-design research, and safety-systems research. Transfer activities include information dissemination, training, technical assistance, and education designed to translate research outputs into content and formats that are designed for application in the workplace. Depending on the scope of the program under review, activities may also be grouped by research-program objectives or subprograms.

Conventional occupational safety and health research focuses appropriately on injury, illness, or death; on biomarkers of exposure; and on health effects of new technology, personal protective equipment, and regulations. A focus on surveillance research may be needed when available data inputs are inadequate. A focus on socioeconomic and policy research and on diffusion research is also needed to effect change, because not all relevant intermediate outcomes occur in the workplace. NIOSH may be able to affect important outcomes farther out on the causal chain so as to influence health and safety in the workplace. Other research that might prove important in addressing NIOSH's mission includes the following:

- Surveillance research to assess the degree of significant or systematic underreporting of relevant injuries, illnesses, and biomarkers
- Socioeconomic research on cost-shifting between worker compensation and private insurance
- Research on methods to build health and safety capacity in community health centers that serve low-income or minority-group workers and to improve recognition and treatment of work-related conditions
- Transfer research to change health and safety knowledge of adolescents while they are in high school to improve the likelihood of reduced injuries as they enter the workforce

TABLE 3 Examples of NIOSH Program Research and Transfer Activities

Surveillance (including hazard and injury, illness, and biomarkers of exposure or effect health surveillance and evaluation of surveillance systems)
Health-effects research (illnesses, injuries, and biomarkers):
Epidemiology
Toxicology
Physical and safety risk factors (laboratory-based)
Development of clinical-screening methods and tools
Exposure-assessment research:
Chemical hazards
Physical hazards
Biologic hazards
Ergonomic hazards
Safety (traumatic injury) hazards
Safer-design and safety-systems research
Intervention research:
Control technologies
Engineering controls and alternatives
Administrative controls
Personal protective equipment
Work organization
Community participation
Policy (such as alternative approaches to targeting inspections)
Design for safety
Emergency preparedness and disaster response
Diffusion and dissemination research:
Training effectiveness
Information-dissemination effectiveness
Diffusion of technology
Health-services and other research:
Access to occupational health care
Infrastructure—delivery of occupational-health services, including international health and safety
Socioeconomic consequences of work-related injuries and illnesses
Worker compensation
Technology-transfer and other transfer activities:
Information dissemination
Training programs
Technical assistance

- Community-based participatory research on differences between recently arrived immigrants and U.S.-born workers regarding perceptions of acceptable health and safety risks so that programs can be targeted to meet the workforce training needs of immigrant workers

Transfer activities should be reviewed to determine whether the NIOSH program appropriately targets its outputs in a manner that will have the greatest impact. Ideally, information dissemination should be proactive, and strategic dissemination should be informed by research on the diffusion of new technologies, processes, and practices. Highly relevant information and technology transfer should include plans for appropriate transfer to all appropriate worker populations, including those considered vulnerable. Training should be incorporated into the strategic goals of all research fields where appropriate.

The EC should review project-level research and transfer activities (including surveillance activities) that have been completed, are in progress, or planned by the program under review. The program under review should provide a list of activities and specify whether they are intramural or extramural. For each extramural project, the key organizations and principal investigators' names should be requested, as should whether the project was in response to a request for proposal or a request for application. For each intramural project, the EC should ask NIOSH to provide a list of key collaborators (from another government agency, academe, industry, or unions).

The EC should evaluate each of the research activities outlined in Table 3 if it forms an important element of the program research. In the case of a sector-based research program (for example, mining or construction) in which health-effects research is not being reviewed, the EC should determine what research inputs influence the program's strategic goals and objective, and then assess the value of the inputs.

Questions to Guide the Evaluation Committee in Assessing Research Activities

1. What are the major subprograms or groupings of activities within the program?
2. Are activities consistent with program goals and objectives?
3. Are research activities relevant to the major challenges of the research program?
   a. Do they address the most serious outcomes?
   b. Do they address the most common outcomes?
   c. Do they address the needs of both sexes, vulnerable working populations, and small businesses?

4. Are research activities appropriately responsive to the input of stakeholders?
5. To what extent are partners involved in the research activities?
6. Are partners involved early in the research process so that they could participate in determining research objectives and research design?
7. Were original resource allocations appropriate for the research activities, and do they remain appropriate?
8. To what extent does peer reviews (internal, external, and midcourse) affect the activities?
9. Is there adequate monitoring of quality-assurance procedures to ensure credible research data, analyses, and conclusions?

Questions to Guide the Evaluation Committee in Assessing Transfer Activities

1. Is there a coherent planned program of transfer activities?
2. Are the program's information dissemination, training, education, technical assistance, or publications successful in reaching the workplace or relevant stakeholders in other settings? How widespread is the response?
3. To what degree have stakeholders responded to NIOSH information and training products?
4. Is there evidence that the formats for information products were selected in response to stakeholder preferences?
5. To what extent do program personnel rely on assessment of stakeholder needs and reactions to prototype information and training projects (formative evaluation techniques)?
6. To what extent does the program build research and education capacity internally and among stakeholders?

Assessment

For this part of the assessment, the EC will provide a qualitative assessment that discusses relevance. This assessment should include consideration of the external factors identified in Section III.A. that constrain choices of research projects and the relevance and effectiveness of transfer activities. The EC should consider the appropriateness of resource allocations. A highly relevant program would address high-priority needs, produce high-quality results, be appropriately collaborative, be of value to stakeholders, and be substantially engaged in transfer activities. A program may be less relevant to the extent that those key elements are not up to the mark or are missing. The discussion should cover those aspects in sufficient detail to arrive at a qualitative assessment of the activities. Assessment of the transfer activities must include considerations of program planning, coherence, and impact. The EC might also consider the incorporation of international research results into NIOSH knowledge-transfer activities for industry sectors in the United States.

III.B.6. Review of Outputs (Figure 2, Box D)

An output is a direct product of a NIOSH research program. Outputs may be designed for researchers, practitioners, intermediaries, and end-users, such as consumers. Outputs can be in the form of publications in peer-reviewed journals, recommendations, reports, Web-site content, workshops and presentations, databases, educational materials, scales and methods, new technologies, patents, technical assistance, and so on. Outputs of the research program's extramurally funded activities should also be considered. Table 4 lists examples of major outputs to be considered by the EC. The NIOSH research program should make every effort to include all pertinent data of the types listed in the table.

Outputs may be tailored to the intended audience to communicate information most effectively and increase the likelihood of comprehension, knowledge, attitude formation, and behavioral intent. The extent of use of formative evaluation data (data gathered before communication for the purpose of improving the likelihood of the intended effects) and the extent of intended user feedback in the design of the output can be considered indicators of appropriate quality assessment.

Some activities such as collaborations can also legitimately be conceptualized as outputs, because the collaboration itself is a result of NIOSH efforts. Cooperation, coordination, more intensive collaboration, and eventual formal partnering can be considered important outputs leading to desirable intermediate outcomes. Technology and knowledge transfer is greatly facilitated through such relationships. The extent of collaboration with other organizations in the determination of research agendas, the conduct of research, the dissemination of research results, and interorganization involvement in the production of outputs can all be measures of output quality and quantity. The EC may consider coauthorship while trying to determine the importance of NIOSH research to the broader research community.

The NIOSH program should provide information on all relevant outputs of the program under review produced during the chosen period.

Questions to Guide the Evaluation Committee

1. What are the major outputs of the research program?
2. Are output levels consistent with resources allocated (were resources allocated and used efficiently to produce outputs)?

TABLE 4 Examples of Research-Program Outputs to Be Considered

Peer-reviewed publications by NIOSH staff:

- Number of original research articles by NIOSH staff
- Number of review articles by NIOSH staff (including best-practices articles)
- Complete citation for each publication
- Complete copies of the "top five" articles
- Collaboration with other public- or private-sector researchers
- Publications in the field of interest with other support by investigators also funded by NIOSH (for example, ergonomic studies with other support by an investigator funded by NIOSH to do ergonomics work, in which case NIOSH should get some credit for seeding interest or drawing people into the field)

Peer-reviewed publications by external researchers funded by NIOSH:

- Number of NIOSH-funded original research articles by external researchers
- Number of NIOSH-funded review articles by external researchers (including best-practices articles)
- Complete citation for each written report
- Complete copies of the "top five" articles
- Collaboration with other government or academic researchers

NIOSH reports in the research program:

- Number of written reports
- Complete citation for each written report
- Complete copies of the "top five" reports

Sponsored conferences and workshops:

- Number of sponsored conferences
- Number of sponsored workshops
- Description of conferences and workshops (title, date, sponsors, target audience, number of participants, and resulting products)
- NIOSH's assessment of value or impact

Databases:

- Number of major databases created by NIOSH staff
- Number of major databases created by external researchers funded by NIOSH grants
- Description of databases:
  - Title, objective (in one to four sentences), and start and stop dates
  - Partial vs. complete sponsorship (if partial, who were cosponsors?)
  - Study or surveillance-system design, study population, and sample size
  - Primary "products" of the database (such as number of peer-reviewed articles and reports)
- Complete copies of the "top two" publications or findings, to date, from each database

continued

TABLE 4 Continued

Recommendations:
Number of major recommendations
Description of recommendations:
Complete citation (article, report, or conference where recommendation was made)
Summary in one to four sentences
Percentage of target audience that has adopted recommendation 1, 5, and 10 years later
Up to three examples of implementation in the field
Identification of "top five" recommendations to date
Tools, methods, and technologies (TMT):
Number of major TMT (includes training and education materials)
Descriptions of TMT
Title and objective of TMT (in one to four sentences)
Complete citation (if applicable)
Percentage of target audience that has used TMT 1, 5, and 10 years later
Up to three examples of implementation in the field
Identification of "top five" TMT to date
Patents:
Total number of patents
For each:
Title and objective (in one to four sentences)
Complete citation
Percentage of target audience that has used product 1, 5, and 10 years later
Up to three examples of implementation in the field
Identification of "top five" patents to date
Miscellaneous:
Any other important program outputs

3. Does the research program produce outputs that address high-priority areas?
4. To what extent does the program generate important new knowledge or technology?
5. Are there widely cited peer-reviewed publications considered to report "breakthrough" results?
6. What, if any, internal or external capacity-building outputs are documented?
7. Are outputs relevant to both sexes, vulnerable populations, and do they address health disparities?
8. Are outputs relevant to health and safety problems of small businesses?

9. Are products user-friendly with respect to readability, simplicity, and design?
10. To what extent does the program help to build the internal or extramural institutional knowledge base?
11. Does the research produce effective cross-agency, cross-institute, or internal-external collaborations?
12. To what extent does the program build research and education capacity (internal or external)?

Assessment

The EC should provide a qualitative assessment discussing relevance and utility. The outputs of a highly ranked program will address needs in high-priority areas, contain new knowledge or technology that is effectively communicated, contribute to capacity-building inside and outside NIOSH, and be relevant to the pertinent populations. The discussion should cover those aspects in sufficient detail to support the qualitative assessment of the outputs.

III.B.7. Review of Intermediate Outcomes (Figure 2, Box E)

Intermediate outcomes are important indicators of stakeholder response to NIOSH outputs. They reflect the impact of program activities and may lead to the desired end outcome of improved workplace safety and health. Intermediate outcomes include the production by those outside of NIOSH of guidelines or regulations based wholly or partly on NIOSH research (products adopted as national or international public policy or as policy or guidelines by private organizations or industry); contributions to training and education programs sponsored by other organizations; use of publications or other materials by workers, industry, and occupational safety and health professionals in the field; and citations of NIOSH research by industrial and academic scientists.

Intermediate outcomes allow inference that a program's outputs are associated with observed changes in the workplace. Thus, an intermediate outcome reflects an assessment of worth by NIOSH stakeholders (such as managers in industrial firms) about NIOSH research or its products (for example, NIOSH training workshops). Intermediate outcomes that are difficult to monitor but may be valid indicators of relevance or utility include self-report measures by users of NIOSH outputs. Such indicators include the extent to which key intermediaries find value in NIOSH products or databases for the repackaging of health and safety information, the extent to which NIOSH recommendations are in place and attended to in workplaces, and employee or employer knowledge of and adherence to NIOSH-recommended practices.

Questions to Guide the Evaluation Committee

1. Do program outputs result in or contribute to stakeholder training or education activities used in the workplace or in school or apprentice programs? If so, how?
2. Do program activities and outputs result in regulations, public policy, or voluntary standards or guidelines that are transferred to or created by the workplace?
3. Has the program resulted in changes in employer or worker practices associated with the reduction of risk (for example, in the adoption of new feasible control or personal protective technologies or administrative control concepts)?
4. Does the program contribute to changes in health-care practices to improve recognition and management of occupational health conditions?
5. Does the program result in research partnerships with stakeholders that lead to changes in the workplace?
6. To what extent do the program's stakeholders find value in NIOSH products (as shown by document requests, Web-site hits, conference attendance, and so on)?
7. Does the program or a subprogram provide unique staff or laboratory capability that is a necessary national resource? If so, is it adequate, or does it need to be enhanced or reduced?
8. Has the program resulted in interventions that protect both sexes, vulnerable workers, or address the needs of small businesses?
9. To what extent did the program contribute to increased capacity at worksites to identify or respond to safety and health threats?

Assessment

Only a qualitative assessment of product development, usefulness, and impact is required at this point in the EC report. Some thought should be given to the relative value of intermediate outcomes, and the FC recommends applying the well-accepted hierarchy-of-controls model. The discussion could include comments on how widely products have been used or programs implemented. The qualitative discussion should be specific as to the various products developed by the program and the extent of their use by specific entities (industry, labor, government, and so on) for specific purposes. Whether the products have resulted in changes in the workplace or in the reduction of risk should be discussed. The recognition accorded to the program or the facilities by its peers (such as recognition as a "center of excellence" by national and international communities) should be considered in the assessment. To be highly ranked, a program should have high performance in most of the relevant questions in this section. An aspect of the evaluation can be whether the same changes in stakeholder activities and behaviors would probably have occurred without NIOSH efforts.

III.B.8. Review of End Outcomes (Figure 2, Box F)

It is necessary for the EC to assess, to the greatest extent possible, NIOSH's contribution to end outcomes—improvements in workplace health and safety (impact). For purposes of this evaluation, end outcomes are health-related changes that are a result of program activities, including decreases in injuries, illnesses, deaths and exposures or risk. Data on reductions in work-related injuries, illnesses, and hazardous exposures will be available for some programs, and in some cases they will be quantifiable. It is possible, however, to evaluate the impact of a NIOSH research program using either intermediate outcomes or end outcomes. If there is no direct evidence of improvements in health and safety, intermediate outcomes may be used as proxies for end outcomes in assessing impact as long as the EC qualifies its findings. The EC will describe the realized or potential benefits of the NIOSH program. Examples of realized intermediate outcomes are new regulations and widely accepted guidelines, work practices, and procedures, all of which may contribute measurably to enhancing health and safety in the workplace.

The FC recognizes that assessing the causal relationship between NIOSH research and specific occupational health and safety outcomes is a major challenge because NIOSH does not have direct responsibility or authority for implementing its research findings in the workplace. Furthermore, the benefits of NIOSH research program outputs can be realized, potential, or limited to the knowledge gained. Studies that conclude with negative results may nevertheless have incorporated excellent science and contribute to the knowledge base. The generation of important knowledge is a recognized form of outcome in the absence of measurable impacts.

The impact of an outcome depends on the existence of a "receptor" for research results, such as a regulatory agency, a professional organization, an employer, and an employee organization. The EC should consider questions related to the various stages that lead to outputs, such as these:

1. Did NIOSH research identify a gap in protection or a means of reducing risk?
2. Did NIOSH convey that information to potential users in a usable form?
3. Were NIOSH research results (for example, recommendations, technologies) applied?
4. Did the applied results lead to desired outcomes?

Quantitative data are preferable to qualitative, but qualitative analysis may be necessary. Sources of quantitative data include the following:

- Bureau of Labor Statistics (BLS) data on fatal occupational injuries (the Census of Fatal Occupational Injuries) and nonfatal occupational injuries and illnesses (the annual Survey of Occupational Injury and Illnesses)
- NIOSH intramural surveillance systems, such as the National Electronic Injury Surveillance System, the coal-worker x-ray surveillance program, and agricultural-worker surveys conducted by NIOSH in collaboration with USDA
- State-based surveillance systems, such as the NIOSH-funded ABLES, and the SENSOR programs (for asthma, pesticides, silicosis, noise-induced hearing loss, dermatitis, and burns)
- Selected state worker-compensation programs
- Exposure data collected in the OSHA Integrated Management Information System

The FC is unaware of mechanisms for surveillance of many occupationally related chronic illnesses, such as cancers that arise from long exposure to chemicals and other stressors. The incidence and prevalence of many such outcomes are best evaluated by investigator-initiated research. Research that leads to new, effective surveillance concepts or programs warrants special recognition.

The EC should recognize the strengths and weaknesses of outcome data sources. Quantitative accident, injury, illness, and employment data and databases are subject to error and bias and should be used by the EC only for drawing inferences after critical evaluation and examination of available corroborating data. For example, it is widely recognized that occupational illnesses are poorly documented in the BLS Survey of Occupational Injuries and Illnesses, which captures only incident cases among active workers. It is difficult for health practitioners to diagnose work-relatedness of most illnesses that may not be exclusively related to work; furthermore, few practitioners are adequately trained to make such an assessment. Many of those illnesses have long latencies and do not appear until years after people have left the employment in question. Surveillance programs may systematically undercount some categories of workers, such as contingent workers.

In addition to measures of illness and injury, measures of exposure to chemical and physical agents and to safety and ergonomic hazards can be useful. Exposure or probability of exposure can serve as an appropriate proxy for disease or injury when a well-described occupational exposure-health association exists. In such instances, a decrease in exposure can be accepted as evidence that the end outcome of reduced illness or injury is being achieved. That is necessary particularly when the latent period between exposure and disease outcome, as in the case of asbestos exposure and lung cancer, makes effective evaluation of the relevant end outcome infeasible.

As an example of how an exposure level can serve as a proxy, reduction in the number of sites that exceed an OSHA PEL or an American Conference of Governmental Industrial Hygienists threshold limit value is a quantitative measure of improvement of occupational health awareness and reduction of risk. In addition to exposure level, the number of people exposed and the distribution of exposure levels are important. Those data are available from multiple databases and studies of exposure. Apart from air monitoring, such measures of exposure as biohazard controls, reduction in requirements for use of personal protective equipment, and reduction in ergonomic risks are important.

Challenges posed by inadequate or inaccurate measurement systems should not drive programs out of difficult fields of study, and the EC will need to be aware of such a possibility. In particular, contingent and informal working arrangements that place workers at greatest risk are also those on which surveillance information is almost totally lacking, so novel methods for measuring impact may be required.

The commitment of industry, labor, and government to health and safety are critical external factors. Several measures of that commitment can be useful for the EC: monetary commitments, attitude, staffing, and surveys of relative importance. To the extent that resources allocated to safety and health are limiting factors, the EC should explicitly assess NIOSH performance in the context of constraints.

Questions to Guide the Evaluation Committee

1. What are the amounts and qualities of relevant end-outcomes data (such as injuries, illness, exposure, and productivity affected by health)?
2. What are the temporal trends in those data?
3. Is there objective evidence of improvement in occupational safety or health?

4. To what degree is the NIOSH program or subprogram responsible for improvement in occupational safety or health?
5. If there is no time trend in the data, how do findings compare with data from other comparable U.S. groups or the corresponding populations in other countries?
6. What is the evidence that external factors have affected outcomes or outcome measures?
7. Has the program been responsible for outcomes outside the United States that have not been described in another category?

Assessment

The EC should provide a qualitative assessment of the program and subprogram impact, discussing the evidence of reductions in injuries and illnesses or their appropriate proxies.

III.B.9. Review of Potential Outcomes

There may be health and safety impacts not yet appreciated and other beneficial social, economic, and environmental outcomes as a result of NIOSH activities. NIOSH study results may be influential outside the United States, and there may be evidence of implementation of NIOSH recommendations and training programs abroad.

Questions to Guide the Evaluation Committee

1. Is the program likely to produce a favorable change that has not yet occurred or not been appreciated?
2. Has the program been responsible for social, economic, security, or environmental outcomes?
3. Has the program's work had an impact on occupational health and safety in other countries?

Assessment

The EC may discuss other outcomes, including beneficial changes that have not yet occurred; social, economic, security, or environmental outcomes; and the impact that NIOSH has had on international occupational safety and health.

III.B.10. Summary Evaluation Ratings and Rationale

The EC should use its expert judgment to rate the relevance and impact of the overall research program by first summarizing its assessments of the major subprograms and then appropriately weighting the subprograms to determine the overall program ratings.

Table 5 provides some background context to aid the EC in reaching overall ratings for relevance and impact. The EC could consider the items in Table 5 for each subprogram then for the overall program and assess the relevance of the research subprograms and program by reviewing earlier responses to the questions in Sections III.B.2. through III.B.5. (reviews of program challenges, strategic goals and objectives, inputs, and activities). Items 1-4 in Table 5 are pertinent to assessing relevance.

To assess overall impact, the EC first needs to consider the available evidence of changes in work-related risks and adverse effects and external factors related to the changes. The EC should review the responses to the questions in Sections III.B.6. through III.B.8. (reviews of outputs, intermediate outcomes, and end outcomes) and systematically assess the impact of the research program and its subprograms. Items 5-7 in Table 5 will be helpful. The EC should evaluate separately the impact of the research and the impact of transfer activities. Transfer activities occur in two contexts: NIOSH efforts to translate intellectual products into practice and stakeholder efforts to integrate NIOSH results into the workplace. High impact assessments require the EC's judgment that the research program has contributed to outcomes; for example, outcomes have occurred earlier than they would have or are better than they would have been in the absence of the research program,

TABLE 5 Background Context for Program Relevance and Impact

Assess the following for each subprogram:

1. Relevance of current and recently completed research and transfer activities to objective improvements in workplace safety and health.
2. Contributions of NIOSH research and transfer activities to changes in work-related practices and reduction in workplace exposures, illnesses, or injuries.
3. Contributions of NIOSH research and transfer activities to improvements in work-related practices.
4. Contributions of NIOSH research to productivity, security, or environmental quality (beneficial side effects).
5. Evidence of reduction of risk in the workplace (intermediate outcome).
6. Evidence of reduction in workplace exposure, illness, or injuries (end outcome).
7. Evidence of external factors that prevented translation of NIOSH research results into intermediate or end outcomes.

or outcomes would have occurred were it not for external factors beyond NIOSH's control or ability to plan around.

The EC must assign one overall integer score for the *relevance* of the research program to the improvement of occupational safety and health and one overall integer score for the *impact* of the program on the improvement of occupational safety and health. The EC will use its expert judgment, summary assessment of research-program elements, and any appropriate information to arrive at those two scores. In light of substantial differences among the types of research programs that will be reviewed and the challenge to arrive at a summative evaluation of both relevance and impact, the FC chose not to construct an algorithm to produce the two final ratings.

Relevance and impact scores will be based on five-point categorical scales established by the FC (see Boxes 2 and 3) in which 1 is the lowest and 5 the highest rating. The FC has made an effort to establish mutually exclusive rating categories in the scales. When the basis of a rating fits more than one category, the highest applicable score should be assigned. It is up to the EC to determine how individual subprograms should influence final scores. Single integer values should be assigned. Final program ratings will consist of integer scores for relevance and impact and prose justification of the scores.

Box 2 includes the criteria for scoring the overall relevance of the NIOSH research program. As discussed in previous sections, numerous factors can be considered in assessing relevance. The scoring criteria focus on two: the EC assessment

BOX 2
Scoring Criteria for Relevance

5 = Research is in high-priority subject areas and NIOSH is significantly engaged in appropriate transfer activities for completed research projects/reported research results.
4 = Research is in priority subject areas and NIOSH is engaged in appropriate transfer activities for completed research projects/reported research results.
3 = Research is in high priority or priority subject areas, but NIOSH is not engaged in appropriate transfer activities; or research focuses on lesser priorities but NIOSH is engaged in appropriate transfer activities.
2 = Research program is focused on lesser priorities and NIOSH is not engaged in or planning some appropriate transfer activities.
1 = Research program is not focused on priorities and NIOSH is not engaged in transfer activities.

BOX 3
Scoring Criteria for Impact

5 = Research program has made major contribution(s) to worker health and safety on the basis of end outcomes or well-accepted intermediate outcomes.
4 = Research program has made some contributions to end outcomes or well-accepted intermediate outcomes.
3 = Research program activities are ongoing and outputs are produced that are likely to result in improvements in worker health and safety (with explanation of why not rated higher). Well accepted outcomes have not been recorded.
2 = Research program activities are ongoing and outputs are produced that may result in new knowledge or technology, but only limited application is expected. Well accepted outcomes have not been recorded.
1 = Research activities and outputs do not result in or are NOT likely to have any application.

of whether the program appropriately sets priorities among research needs and the EC assessment of how engaged the program is in appropriate transfer activities. Table 6 provides some guidance regarding how the EC may weight research priorities and transfer levels when determining relevance scores.

The EC will consider both completed research and research that is in progress and related to likely future improvements in its assessment of relevance. The EC should keep in mind how well the program has considered the frequency and severity of the problems being addressed; whether appropriate attention has been directed to both sexes, vulnerable populations, or hard-to-reach workplaces; and whether the different needs of large and small businesses have been accounted for. It is up to the EC to determine how to consider external factors in assigning program scores.

Box 3 includes the criteria established for the rating of impact. In general, the EC will consider completed research outputs during the assessment of impact. In assigning a score for impact, it is important to recognize that a "major contribution" (required for a score of 5) does not imply that the NIOSH program was solely responsible for observed improvements in worker health and safety. Many factors may be required to effect improvements. The EC could say that NIOSH made "major contributions" if the improvements would not have occurred when they did without NIOSH efforts.

TABLE 6 Guidance for Weighting Research Priority and Engagement in Appropriate Transfer Activities in the Application of Relevance Score

Assessment of Research Priority	Engagement in Applicable Transfer Activities	Applicable Score
High priority	Significantly engaged	5
High priority	Engaged	4
High priority	Not engaged	3
Priority	Significantly engaged	4
Priority	Engaged	4
Priority	Not engaged	3
Lesser priority	Significantly engaged	3
Lesser priority	Engaged	3
Lesser priority	Not engaged	2
Not focused on priorities	Significantly engaged	2
Not focused on priorities	Engaged	2
Not focused on priorities	Not engaged	1

The FC has some concern that the imposed scoring criteria for impact might be considered a promotion of the conventional occupational-health research paradigm that focuses on health-effects and technology research without much emphasis on the socioeconomic, policy, surveillance, and diffusion research (as opposed to diffusion activities) needed to effect change. The EC should remember that not all intermediate outcomes occur in the workplace. Important outcomes that NIOSH can effect also occur much farther out on the causal chain. NIOSH, for example, has an important role to play in generating knowledge that may contribute to changing norms in the insurance industry, in health-care practice, in public-health practice, and in the community at large. The EC may find that some of those issues need to be addressed and considered as external factors that facilitate or limit application of more traditional research findings. Given the rapidly changing nature of work and the workforce and some of the intractable problems in manufacturing, mining, and some other fields, the EC is encouraged to think beyond the traditional paradigm.

III.C. Assessment of NIOSH Process for Targeting Priority Research Needs and Committee Assessment of Emerging Issues

The second charge to the EC is the assessment of the research program's effectiveness in targeting new research and identifying emerging issues in occupational safety and health most relevant to future improvements in workplace protection. The EC is also asked to provide a qualitative narrative assessment of the program's efforts and to make suggestions about emerging issues that the program should be prepared to address. Among the most challenging aspects of research in illness and injury prevention are the identification of new or emerging needs or trends and the formulation of a research response that appropriately uses scarce resources in anticipation of them.

The EC should review the procedures that NIOSH and the research program have in place to identify needed research relevant to the NIOSH mission and should review the success that NIOSH has had in identifying and addressing research related to emerging issues. It should examine leading indicators from appropriate federal agencies, such as EPA, the Department of Labor, the National Institute of Standards and Technology, NIH, DOD, and the Department of Commerce. Those indicators should track new technologies, new products, new processes, and disease or injury trends.

One source of information deserving particular attention is NIOSH HHE reports. The HHE program offers a potential mechanism for identifying emerging research needs that could be incorporated as input into each of the programs evaluated. The EC should determine whether the program under review appropriately considers pertinent HHE investigation findings. Additional emerging issues may be revealed through consideration of NIOSH and the NIOSH-funded FACE reports, the AOEC reports, the U.S. Chemical Safety Board investigations, and SENSOR and other state-based surveillance programs. Appropriate federal advisory committees and other stakeholder groups should also be consulted to provide qualitative information.

The EC should systematically assess how the research program and its subprograms target new research by evaluating each subprogram for the items listed in Table 7. The EC will have to determine how best to weight subprogram contributions in the program's targeting of new research.

TABLE 7 Targeting of New Research and Identification of Emerging Issues

Assess the following for each subprogram:

1. Past and present effectiveness in targeting most relevant research needs.
2. Effectiveness in targeting research in fields most relevant to future improvements in occupational safety and health.
3. Contribution of NIOSH research to enhancement of capacity in government or other research institutions.

Questions to Guide the Evaluation Committee

1. What information does NIOSH review to identify emerging research needs?
   a. What is the process for review?
   b. How often does the process take place?
   c. How are NIOSH staff scientists and NIOSH leadership engaged?
   d. What is the process for moving from ideas to formal planning and resource allocation?
2. How are stakeholders involved?
   a. What advisory or stakeholder groups are asked to identify emerging research targets?
   b. How often are such groups consulted, and how are suggestions followed up?
3. What new research targets have been identified for future development in the program under evaluation?
   a. How were they identified?
   b. Were lessons that could help to identify other emerging issues learned?
   c. Does the EC agree with the issues identified and selected as important and with the NIOSH response, or were important issues overlooked?
   d. Is there evidence of unwise expenditure of resources on unimportant issues?

The EC members should use their expert judgment both to evaluate the emerging research targets identified by NIOSH and to provide recommendations to NIOSH regarding additional research that NIOSH has not yet identified. Recommendations should include a brief statement of their rationale.

IV. EVALUATION COMMITTEE REPORT TEMPLATE

Consistency and comparability among EC report formats is desirable, but the FC recognizes that each NIOSH research program is different and that each EC is independent. The outline provided in Box 4 flows from the FC's review of NIOSH's generalized logic model (Figure 1), the evaluation flowchart (Figure 2), and the assessment model described earlier in this document. The EC should feel free to use or adapt this outline as necessary when organizing its final report. The FC encourages each EC to look at prior EC reports for organizational ideas.

BOX 4
Suggested Outline for Evaluation Committee Reports

I. Introduction

This section should be a brief descriptive summary of the history of the program (and subprograms) being evaluated with respect to pre-NORA, NORA 1, and current and future plans of the research program presented by NIOSH. It should present the context for the research on safety and health; goals, objectives, and resources; groupings of subprograms; and any other important pertinent information. (A list of the NIOSH materials reviewed should be provided in Appendix C.)

II. Evaluation of Programs and Subprograms (Charge 1)

A. Evaluation summary (should include a brief summary of the evaluation with respect to impact and relevance, scores for impact and relevance, and summary statements).
B. Strategic goals and objectives: should describe assessment of the program and subprograms for relevance.
C. Review of inputs: should describe adequacy of inputs to achieve goals.
D. Review of activities: should describe assessment of the relevance of the activities.
E. Review of research-program outputs: should describe assessment of relevance and potential usefulness of the research program.
F. Review of intermediate outcomes and causal impact: should describe assessment of the intermediate outcomes and the attribution to NIOSH; should include the likely impacts and recent outcomes in the assessment.
G. Review of end outcomes: should describe the end outcomes related to health and safety and provides an assessment of the type and degree of attribution to NIOSH.
H. Review of other outcomes: should discuss health and safety impacts that have not yet occurred; beneficial social, economic, and environmental outcomes; and international dimensions and outcomes.
I. Summary of ratings and rationale.

III. NIOSH Targeting of New Research and Identification of Emerging Issues (Charge 2)

The EC should assess the progress that the NIOSH program has made in targeting new research in occupational safety and health. The EC should assess whether the NIOSH program has identified important emerging issues that appear especially important in terms of relevance to the mission of NIOSH. The EC should respond to NIOSH's perspective and add its own recommendations.

IV. Recommendations for Program Improvement

On the basis of the review and evaluation of the program, the EC may provide recommendations for improving the relevance of the NIOSH research program to health and safety conditions in the workplace and the impact of the research program on health and safety in the workplace.

Appendix A — Framework Document
Appendix B — Methods and Information-Gathering
Appendix C — List of NIOSH and Related Materials Collected in the Process of the Evaluation

B

Committee Meetings

The Committee to Review the NIOSH Construction Research Program held three face-to-face meetings during the course of its study. The first two meetings included open sessions for information gathering. The agendas for these open sessions appear below. The third meeting was held in closed session.

MEETING I

July 17, 2007
The Keck Center of the National Academies
Washington, D.C.

Tuesday, July 17, 2007

12:15-4:45 p.m. Construction Program Overview
Lewis Wade, NIOSH

National Construction Center Overview,
The Center to Protect Workers' Rights (CPWR)
Frank Hearl, NIOSH Program Manager
Peter Stafford, CPWR Executive Director

Goal 1: Prevent Injuries and Fatalities
David Fosbroke, NIOSH Program Staff
Hester Lipscomb, Duke University Medical Center

Goal 2: Reduce Overexposures to Health Hazards
Scott Earnest, NIOSH Program Staff
Mark Goldberg, City University of New York

Break

Goal 3: Reduce Musculoskeletal Disorders
Laura Welch, CPWR
Renguang Dong, NIOSH Program Staff

Goal 4: Increase Understanding of Factors for Improving Outcomes
Janie Gittleman, CPWR
Marie Haring-Sweeney, NIOSH Program Staff

Future Directions
Matt Gillen, NIOSH Program Staff

4:45–5:15 p.m. Public Comment

5:30 p.m. Open session adjourned

MEETING II

September 25-26, 2007
The Keck Center of the National Academies
Washington, D.C.

Tuesday, September 25, 2007

11:30 a.m. NIOSH Construction Program Team
- Project Table
- Site Visit

1:15 p.m.	Panel 1: Contractors/Owners • *Gary Fore, National Asphalt Paving Association* • *Bill Isokait, American Subcontractors Association* • *Robert Laramore, American Road and Transportation Builders Association* • *Kevin Cannon, National Association of Home Builders*
3:15 p.m.	Panel 2: Regulatory/Government • *Daniel Murphy, Zurich North America* • *Stewart Burkhammer, Occupational Safety and Health Administration (OSHA) Directorate of Construction* • *David Valiante, New Jersey Department of Health and Senior Services, Occupational Health Service* • *Charles Stribling, Kentucky Department of Labor*
5:00 p.m.	Public Comment
5:30 p.m.	Committee recesses
6:00 p.m.	Working discussions over dinner (committee members only)

Wednesday, September 26, 2007

8:45 a.m.	Panel 3: Worker/Labor Management Organizations • *William Hering, Association of Union Constructors* • *Frank Migliaccio, Iron Workers International Union* • *Pablo Alvarado, National Day Laborer Organizing Network* • *Peter Dooley, National Council for Occupational Safety and Health* • *Travis Parsons, Laborers' Health and Safety Fund of North America*
10:30 a.m.	Break
10:45 a.m.	Panel 4: Occupational Health and Safety • *Steven Fess, American Industrial Hygiene Association* • *Daniel Murphy, Construction Safety Council*

- *Tee Guidotti, American College of Occupational and Environmental Medicine and Association of Occupational and Environmental Clinics*
- *David Heidorn, American Society of Safety Engineers*

12:30 p.m. Public Comment
- *Don Ellisburg, Consultant*

1:00 p.m. Lunch

1:45 p.m. NIOSH Construction Program Team
- Issue review/summary with committee

2:30 p.m. Open session adjourned

MEETING III

December 10-12, 2007
The Beckman Center of the National Academies
Irvine, California

Closed Session

C

Board on Infrastructure and the Constructed Environment

DAVID J. NASH, *Chair*, Dave Nash & Associates, Washington, D.C.
JESUS DE LA GARZA, Virginia Tech, Blacksburg
REGINALD DesROCHES, Georgia Institute of Technology, Atlanta
DENNIS DUNNE, dddunne & associates, Scottsdale, Arizona
BRIAN ESTES, U.S. Navy (retired), Williamsburg, Virginia
PAUL FISETTE, University of Massachusetts, Amherst
LUCIA GARSYS, Hillsborough County, Florida
THEODORE C. KENNEDY, BE&K, Inc., Birmingham, Alabama
PETER MARSHALL, Dewberry Company, Norfolk, Virginia
DEREK PARKER, Anshen+Allen Architects, San Francisco, California
JAMES PORTER, DuPont, Wilmington, Delaware
E. SARAH SLAUGHTER, Massachusetts Institute of Technology, Cambridge
WILLIAM WALLACE, Rensselaer Polytechnic Institute, Troy, New York

Staff

GARY FISCHMAN, Director
LYNDA STANLEY, Senior Program Officer
KEVIN LEWIS, Program Officer
HEATHER LOZOWSKI, Financial Associate
TERI THOROWGOOD, Administrative Coordinator
LAURA TOTH, Program Assistant

D

Biosketches of Committee Members

Richard L. Tucker (*Chair*) is the Joe C. Walter, Jr., Chair in Engineering Emeritus at the University of Texas, Austin. He currently serves on the board of directors for Hill and Wilkinson, Inc., Integrated Electrical Services. He is a member of the National Academy of Engineering, a fellow of the American Society of Civil Engineers, and a member of numerous professional societies and associations, including the National Society of Professional Engineers, the American Association of Cost Engineers, and the American Society for Testing and Materials. Dr. Tucker currently serves on the Committee on the Review of NIOSH Research Programs, known as the Framework Committee. His awards and honors include the Construction Engineering Educator Award of the National Society of Professional Engineers, the Construction Industry Institute's Ronald Reagan Award for Individual Initiative and Carroll H. Dunn Award, and the Michael Scott Endowed Research Fellow of the Institute for Constructive Capitalism. He has published numerous items spanning four decades, and wrote "Communicating in Construction: The Path to Project Success," Chapter 1 of the *1996 Wiley Construction Law Update.* Dr. Tucker has a B.S., an M.S., and a Ph.D. in civil engineering from the University of Texas at Austin. His areas of expertise include project management aspects of capital facilities delivery, construction project planning, construction productivity, improving efficiency and effectiveness of design and procurement, and contracting.

Paul Barshop is the chief operating officer of Independent Project Analysis (IPA), a consulting firm that specializes in construction project evaluation, benchmarking, and analysis. He joined IPA in 1994 as a project analyst and was IPA's quality manager from 1997 to 1999. From 2000 until mid-2004, he was the director of IPA's Netherlands office with the responsibility of serving clients in Europe, the Middle East, and Africa. As a project analyst, Mr. Barshop focused on evaluating downstream process projects, especially in the petroleum and chemical areas. He led numerous benchmarking efforts and conducted more than 75 individual analyses of capital projects. He also led research to contribute to the understanding of the performance and drivers of control system projects. His latest research efforts include the study of the effectiveness of engineering value centers and the study of best practices. Mr. Barshop holds a master's degree in business and a bachelor's degree in chemical engineering. Prior to joining IPA, he worked for Shell Oil Company in the United States. His areas of expertise include benchmarking and best practices in the construction industry.

Maria Brunette is an assistant professor in the Department of Work Environment at the University of Massachusetts, Lowell. She has published numerous articles and conducted research across a broad spectrum of topics, including quality of work life, occupational stress, and the safety and health of Hispanic workers in the United States. Dr. Brunette's work focuses on applying human factors and systems engineering to the design of work systems. Her interests include methods for measuring the role of job, organizational, and cultural factors in the quality of work life. In all of these areas, her focus is on underrepresented ethnic groups and women, especially those of Hispanic origin. Her scholarly interests include macroergonomics, job and organizational design, job satisfaction and stress, and occupational safety and health. Dr. Brunette received her degrees in industrial engineering from the University of Lima, Peru (B.Sc.), from the University of Puerto Rico, Mayaguez (M.Sc.), and from the University of Wisconsin-Madison (Ph.D.). Her areas of expertise include ergonomics, human factors and engineering, and Hispanic workers.

Patricia A. Buffler is professor of epidemiology and holds the Kenneth and Marjorie Kaiser Chair in Cancer Epidemiology in the School of Public Health at University of California, Berkeley. Dr. Buffler's research interests include the environmental causes of cancer, especially gene-environment interaction and childhood cancer, lung cancer, leukemia, brain cancer, and breast cancer; epidemiologic research methods; and the uses of epidemiologic data in health policy. She has served on numerous committees of the National Research Council including the following: Committee on Gulf War and Health: Health Effects Associated with Exposure During the Persian Gulf War; Subcommittee to Review the Hanford Thyroid Disease Study Final Results and Report; Committee on Environmental Justice: Research,

Education, and Health Policy Needs; National Forum on Science and Technology Goals: Environment; Steering Committee on Valuing Health Risks, Costs, and Benefits for Environmental Decisions; Committee on Chemical Toxicity and Aging; Committee on Passive Smoking; Committee on Non-occupational Health Risks of Asbestiform Fibers; and Committee on Research of Agents Potentially Hazardous to Human Health. She also served on the Committee to Assess Health Risks from Exposure to Low Levels of Ionizing Radiation (BEIR [Biological Effects of Ionizing Radiation] VII Phase 2). Dr. Buffler was elected to the Institute of Medicine in 1994. She received a Ph.D. in epidemiology from the University of California, Berkeley.

Angela DiDomenico is a researcher for the Liberty Mutual Research Institute for Safety. Her areas of expertise and research include postural stability and fall prevention, occupational biomechanics, and workload assessment. Currently she conducts research within the slips-and-falls domain and seeks to determine the factors that significantly cause or contribute to workplace injuries. Her recent investigations involve measuring postural stability during various tasks, including while working at elevations and following postural transitions to a standing posture. Prior to joining the staff at Liberty Mutual, Dr. DiDomenico served as a research assistant in the Industrial Ergonomics Laboratory at Virginia Tech, Blacksburg. At the university, she earned her Ph.D. and M.S. in industrial and systems engineering (human factors) and her M.S. in mathematics. While completing her studies, she received a number of awards and honors, including an internship with the Army Research Laboratory (Aberdeen, Maryland), the Thompson Scholarship for Women in Safety presented by the American Society of Safety Engineers Foundation, and a fellowship from the National Institute for Occupational Safety and Health.

Bradley Evanoff is chief of the Division of General Medical Sciences in the Department of Internal Medicine at Washington University. He also heads the Section of Occupational and Environmental Medicine and holds the Richard A. and Elizabeth Henby Sutter Chair in Occupational, Industrial, and Environmental Medicine. Dr. Evanoff's areas of expertise include the epidemiology and prevention of work-related musculoskeletal disorders, work-related health problems in health care workers, and the evaluation of occupational medical education. In addition to his academic duties, Dr. Evanoff is involved in many issues related to employee health and safety at Barnes-Jewish Center Healthcare. He received his undergraduate degree from Cornell University and his medical degree from Washington University. Following a residency in internal medicine at Barnes-Jewish Hospital, he completed a Fogarty postdoctoral fellowship at the Swedish National Institute of Occupational Health. He was then a fellow in the Robert Wood Johnson Clinical Scholars Program and the Occupational and Environmental Medicine Program at the University of Washington in Seattle, where he also received his master's degree in public health. Dr. Evanoff's areas of expertise also include occupational and environmental medicine, epidemiology, the design and analysis of intervention studies, and the use of functional status measures.

Linda M. Goldenhar is currently an assistant dean, associate professor of family medicine, and director of the Office of Evaluation and Research in Medical Education at the University of Cincinnati College of Medicine. She received her Ph.D. from the University of Michigan School of Public Health. For 9 years prior to joining the University of Cincinnati, she was a research psychologist and team leader of the Intervention Effectiveness Research National Occupational Research Agenda (NORA) team at the National Institute for Occupational Safety and Health. Dr. Goldenhar's areas of expertise include both qualitative and quantitative research methods to conduct program evaluation and research studies. Her evaluations have covered a variety of content areas including the safety and health of women working in nontraditional occupations, job stress, and, in particular, intervention evaluation in occupational health and safety as well as medical education. She is widely published and has been invited to present both nationally and internationally on these topics. She is an associate editor of *Public Health Reports,* is on the editorial board of the *Journal of Safety Research,* and is a regular reviewer for *Work and Stress* and the *Journal of Occupational Health Psychology.*

William H. Kojola is the industrial hygienist for the AFL-CIO. His experience in health and safety spans more than 25 years, during which he has been the director of the Occupational Safety and Health Division of the Laborers Health and Safety Fund of North America, an occupational safety and health specialist for the International Brotherhood of Boilermakers, and director of safety and health for the United Cement, Lime, Gypsum and Allied Workers International Union. Prior to this, Mr. Kojola was a health research scientist at the University of Illinois School of Public Health, studying the human health effects of air and water pollutants. With the AFL-CIO, he is responsible for developing strategies for securing new safety and health protections through federal and state regulations, coordinating with affiliates on and leading a unified labor response to proposed Occupational Safety and Health Administration regulations, and representing the AFL-CIO before government regulatory agencies, on federal advisory committees, and in consensus standard-setting efforts. He also provides technical and strategic support to organizing campaigns on safety and health issues. Mr. Kojola holds a B.S. degree in biology and an M.S. degree in genetics from the University of Minnesota and studied toxicology and industrial hygiene at the University of Illinois School of Public Health. His areas of expertise include industrial hygiene and labor relations.

Emmitt J. Nelson is president of Nelson Consulting, Inc., a safety consulting firm specializing in the Zero Injury Concept in eliminating worker injury. He was retained by Shell Oil Company in 1990 to chair the "Zero Accidents Task Force" of the Construction Industry Institute (CII) located at the University of Texas. Through research, this task force identified the unique safety management techniques that successful contractors and owners use in the construction industry to achieve zero worker injuries. Mr. Nelson has continued to analyze the research data and uses this information to lead his clients into new realms of excellence in safety performance. He is a graduate of Texas A&M University with a degree in mechanical engineering. He has been honored with the Business Roundtable Construction Safety Excellence Award; was named CII Co-Instructor of the Year in 2003, and was inducted into the National Academy of Construction on the basis of his contribution to safety in the construction industry. He is a licensed professional engineer in the state of Texas and an associate member of American Society of Safety Engineers. His areas of expertise include the construction industry and construction safety practices.

Peter Philips is professor and chair of the economics department at the University of Utah. He received his B.A from Pomona College in 1970 and his Ph.D. from Stanford University in 1980. He is a labor economist and economic historian. His research focuses on the construction industry. Dr. Philips's areas of expertise include prevailing wage laws and employment, training, wages, benefits, and safety in the construction industry. He has many academic publications to his credit. He has also served as an expert on the construction industry for the U.S. Labor Department and the U.S. Department of Justice. Dr. Philips has testified before many state legislatures on construction regulation issues. His most recent books, *Building Chaos: An International Comparison of Deregulation in the Construction Industry* (Routledge Press, 2003) and *The Economics of Prevailing Wage Laws* (Ashgate Press, 2005), and his most recent journal articles focus on school construction costs, construction labor market regulation, fatalities in the construction workplace, and the effect of subcontracting on construction safety. He is co-editor with Garth Mangum of *Three Worlds of Labor Economics* (M.E. Sharpe, 1986) and coauthor of *Portable Pensions for Casual Labor Markets: The Central Pension Fund of the Operating Engineers* (Quorum Books, 1995).

Iris D. Tommelein is professor of engineering and project management in the Civil and Environmental Engineering Department at the University of California, Berkeley. She teaches and conducts research on what is termed lean construction, developing the theory and principles of project-based production management for the architecture-engineering-construction industry. Professor Tommelein's current research focuses on the work specialty that contractors and suppliers perform and how they can become integral participants in design-build teams in order to increase process and product development performance. Her areas of expertise include construction site logistics, layout, materials management, supply-chain management, and electronic commerce. Her work involves computer-aided design, planning, scheduling, simulation, and visualization of construction processes; and the use of information technology including Web-based systems, wireless communication, bar coding, and laser-based positioning systems. Professor Tommelein is the executive director of the newly formed Production Systems Laboratory, a research institute dedicated to developing and deploying knowledge and tools for project management as well as a learning laboratory for the Northern California construction industry. She is an active participant in the International Group for Lean Construction, and she serves on the board of directors of the Lean Construction Institute. Professor Tommelein served on the executive committee of the American Society of Civil Engineers (ASCE) Technical Council on Computing and Information Technology. She is a member of the Construction Research Council of the Construction Institute of the ASCE. She is the 2002 recipient of the Walter L. Huber Civil Engineering Prize for her research on civil engineering computing for managing project-based production systems in the engineering-architecture-construction industry.